PRACTICAL HEAT RECOVERY

John L. Boyen

PRACTICAL HEAT RECOVERY

A Wiley-Interscience Publication

JOHN WILEY & SONS

New York/London/Sydney/Toronto

Copyright © 1975, by John Wiley & Sons, Inc.

Library of Congress Cataloging in Publication Data:

Boyen, John L   1911–
  Practical heat recovery.

  "A Wiley-Interscience publication."
  Bibliography: p.
  Includes index.
  1. Heat recovery.   I. Title.

TJ260.B67   621.4′02   75–17735
ISBN 0–471–09376–9

Printed in the United States of America

10 9 8 7 6 5 4 3 2 1

PREFACE —————————————————————————

This book is a product of many years of working in the field of heat recovery, as well as the requirement to use many different reference sources for data. For this reason, I have attempted in this book to coordinate the data, mostly in the convenient form of curves, tables, and charts to make the solution of design problems in heat recovery easier to solve. The methods employed have been repeatedly used in practice, resulting in equipment that met or slightly exceeded contract requirements, thereby verifying the procedure.

The future promises ever-increasing fuel costs and shortages. Therefore, the requirement and incentives for heat recovery will be ever more important. Among the more obvious applications for heat recovery are gas turbine exhaust, reciprocating engine exhaust, and incinerator exhaust. For this reason, I have stressed these three forms, and included air-to-air heat recovery, air-to-gas, gas-to-water, gas-to-organic fluids, and gas-to-steam generation. The gas-to-steam section includes steam for heating, air conditioning, processing, and power generation. One of the most effective forms in the steam generation phase is steam generation by waste heat recovery, used for supplementary power generation in the combined gas turbine–steam turbine power cycle. This power cycle is thoroughly discussed in this book.

The final chapter is devoted to the proper application of insulating material to adequately conserve heat for the process and limit surface temperatures for safety reasons.

The book is directed to engineers in the field of waste heat conservation, consulting engineers, students of heat transfer and mechanical engineering, and potential users of heat-recovery equipment. As it is assumed that the reader already has a working knowledge of the heat-transfer processes, the book does not dwell on theory to any great degree except as a basis for the practical aids presented.

It is hoped that the material included will be helpful to those interested in the potentials for practical heat recovery, and will provide a frame of reference for the additional information which will continue to be developed in this field.

JOHN L. BOYEN

Emeryville, California
March 1975

CONTENTS ─────────────────────────────────

INTRODUCTION ————————————

OBJECTIVE

This treatise was written to assist engineers involved in the design and specification of waste heat recovery equipment. The convenient methods of determining the characteristics of heat recovery equipment are intended to avoid complex computations.

SCOPE

Heat recovery from waste gas sources—gas turbine, gas, and diesel engines and the incineration of waste gas, liquids or solids—presents an ever increasing opportunity for economical operation of thermal systems. The actual economic advantage from any form of heat recovery depends on the availability and cost of fuels. Obviously, savings from heat recovery increase as fuel costs rise. The cost of fuel saved must be compared to the capital investment, amortization, maintenance, operating cost, taxes, insurance, and any other cost factors involved in owning and operating the equipment. The payback period must be determined for each case. However, the following payback periods are typical:

1. Boiler or incinerator air-to-air heat recovery—2 to 3 years
2. Heat recovery liquid heater on incinerator or engine exhaust—3 to 4 years
3. Heat recovery boiler (low pressure) on gas turbine or incinerator exhaust—4 to 5 years
4. Heat recovery boiler (high pressure) on gas turbine or incinerator exhaust—5 to 7 years

A simple method of estimating return on investment for a specific case (air-gas heat recovery) is outlined in Chapter 2. The same general method may be used for any other forms of heat recovery.

1

The heat recovery process itself does not reduce air or water pollution except for the small percentage of airborne or gas-borne solid particles that adhere to the working surface of the equipment. Combustion equipment such as various forms of incineration does, however, reduce air or water pollution, and this is its primary purpose. Heat recovery equipment is intended to recover only as much of the heat of combustion as practical, in the form most useful to the particular plant process. Heat recovery equipment used in conjunction with incinerators, particularly solids or liquids incinerators, is always followed by scrubbers and cyclones to remove solids from the gas stream to meet the requirements of government regulating agencies. A discussion of the design and selection of such scrubber and solids removal equipment is not a part of this treatise, since the design of such equipment deserves separate treatment.

Combustion products of the various heat sources vary with the fuel input and the characteristics and composition of the incinerated waste material. Consideration of these characteristics is important to the design and selection of heat recovery equipment only as regards possible corrosive effects of noxious elements in the combustion products.

The percentages by weight of combustion products from a typical sewage sludge incinerator are listed as follows:

Carbon dioxide—12%
Nitrogen—65%
Oxygen—9%
Water vapor—14%
Ash—about 4 lb per 1000 cu ft of gas

Sometimes there is a small amount of SO_2 present, which is removed in the final clean up. As is evident, however, the major contaminant is ash. Most of the ash from an incinerator is removed directly by ash-handling conveyors. We are concerned here with the ash that is carried into the hot gas stream and into the heat recovery equipment. Some of the ash in the gas stream adheres to the heating surfaces and must be periodically removed by air or steam-soot blowing. A portion falls to the bottom of the incinerator and can be removed by ash-handling hoppers and conveyors or by other means. The remaining ash is carried through with the gas stream and must be removed by wet cyclones, scrubbers, and the like.

Maximum permissible quantities of some common pollutants from waste combustion according to the October 1972 issue of Occupational Safety and Health Administration regulations are

Sulfur dioxide—5 ppm
Nitric oxide—25 ppm

Fluoride—2.5 ppm
Hydrogen bromide—3.0 ppm
Ammonia—50 ppm

These regulations must be consulted in designing pollution abatement equipment for a particular case. Each case, being different, must be considered separately. However, the design of pollution abatement equipment and systems is worthy of separate consideration and is not the primary subject of this treatise.

We will discuss the recovery of heat from gas turbine, gas, or diesel engines and from incinerators for waste gas, liquids, or solids.

In any of these applications, heat exchange can take several forms: air-to-gas, gas-to-water, gas-to-organic fluids, gas-to-water, and steam. In the last example, steam can be generated at low pressure for heating or absorption air-conditioning applications, at medium pressures (100 to 150 psig) for processing, or at higher pressures (250 to 600 psig) with or without superheat for power generation. A chapter will be devoted to each mode of heat recovery.

The selection of the mode of heat recovery depends on the characteristics of the application, the processes used by the particular facility, and the economic need for a given service. For example:

1. A gas turbine in natural gas compressor service is ordinarily in remote locations in an automatically controlled facility requiring no hot water, steam, or heat transfer fluid. Such an application has only one remaining usable fluid—air. An air-to-exhaust gas recuperator can be used advantageously to reduce turbine fuel rate.

2. A sewage sludge incinerator with an afterburner to eliminate unburned hydrocarbons from its exhaust gas flow, located where steam generation is not required, can use a gas-to-combustion air recuperator to improve efficiency of afterburning. A gas-to-gas heat exchanger can be used to reheat scrubber exhaust for plume suppression.

3. A gas turbine or reciprocating engine installation driving an alternator in a shopping center, hospital, school, or other commercial or industrial site as part of a total energy system can effectively use exhaust heat recovery in the form of medium temperature hot water or low pressure steam for use in an absorption chiller system for air conditioning.

4. A gas turbine driving a centrifugal refrigeration compressor in a central heating and cooling plant for a building complex can recover exhaust heat in the form of high pressure steam (250 psig and higher) to drive a steam-turbine-driven refrigeration compressor to provide additional refrigeration compressor to provide additional refrigeration capacity.

5. In a sewage sludge heat treatment facility, treated and dewatered sludge

can be burned in a multiple-hearth incinerator. The gas temperature leaving the afterburner section can range from 1000 to 1400°F. Heat from this source can be recovered by high pressure steam generation (250 to 275 psig) for use in the sludge heat treatment process, eliminating a fired boiler for this purpose. If more heat is recovered than can be used in the sludge heat treatment process, the surplus steam can be used in space heating at reduced pressure.

6. Plastics plants produce a substantial amount of liquid waste containing combustible material. These plants also use heated organic heat transfer fluids in producing plastics in heated retorts, kettles, continuous heat exchangers, and so on. In such plants, it is economical to burn the waste liquid in an incinerator and recover the heat as hot fluid (Dowtherm "A" or other fluid) in a heat recovery liquid-phase heater or Dowtherm boiler.

7. Gas turbines are now common on offshore oil production platforms. They provide electrical power for the platform and drive natural gas compressors and high pressure water pumps for well stimulation. The exhaust gas heat from these turbines can be used to heat ethylene glycol for heating crew's quarters or to heat an organic heat transfer fluid for crude oil heating by the use of liquid-phase heat recovery units.

8. Electrical utilities are now adding gas-turbine-based equipment to their conventional steam plants, using one or more gas turbines driving alternators. The exhaust gas from the gas turbines is used to generate high pressure steam (at 650 psig and more) to drive a steam-turbine-driven alternator. Economizers and superheaters enhance cycle efficiency. The lower temperature gas from the economizer then generates low pressure steam (5 to 10 psig) for feedwater heating and deaerating. The result is an overall efficiency approaching 40%. The overall cycle efficiency of a conventional electrical utility steam plant is 32 to 34%. This has become known as the combined cycle system plant and is an excellent example of the benefits of efficient heat recovery.

In addition to the aforementioned applications, heat recovery is provided by open-hearth furnaces, H_2S combustion, refining furnaces, pyrite roasters, petrochemical cracking furnaces, ammonia combustion, sulfur combustion, zinc recovery, and others. The mode of heat recovery can be steam generation, steam superheating, water heating, organic heat transfer fluid heating, and gas-to-air or air-to-air heat exchange. The mode must be appropriate to each application, depending upon the requirements of the process, as previously stated. The kind of equipment discussed in the text lends itself to a range of applications. Usually only the material of construction has to be selected to satisfy pressure, temperature, and corrosion conditions.

There is a great deal of literature on the design of liquid-to-liquid, steam-to-liquid, and fluid-to-vapor heat exchange equipment; such equipment can be included in the heat recovery category because it is frequently used in heat recovery applications. This equipment is broadly classified as shell-and-tube heat exchangers. Since the methods and principles of its design and application are so well known, it is not treated here.

MECHANICS OF HEAT TRANSFER

We assume that the reader is already familiar with the basic principles of heat transfer and thermodynamics. However, the following review is probably in order.

Heat transfer may be broken down into three basic mechanisms: (1) conduction, (2) convection, and (3) radiation.

Conduction is the transfer of heat from one molecule or particle of matter to another, the molecules or particles remaining in fixed positions relative to each other.

Calculation of the rate of heat conduction through any material involves measuring the rate of heat flow through the material by conduction per unit of cross-sectional area taken normal to the direction of heat flow and per unit temperature gradient measured in the direction of heat flow.

Thermal conductivities of different materials vary widely. Some common values are listed in the Appendix.

The unit of thermal conductivity commonly used is $Btu/(ft)(h)(F°)$.

Conduction of heat through any body is unsteady if the temperature at any point in the body varies with time; conduction is steady if the temperature at every point in the body remains constant. Hence, with unsteady conduction the temperature gradient dt/dL at each point in the body also varies with time. As an example, take an insulated pipe. When steam first flows through the pipe, the temperature at the inner surface of the insulation rises very rapidly, but the temperature at any point near the outer surface is not immediately affected. The temperature at each point in the insulation continues to rise for some time, which is unsteady conduction. Finally, after the system reaches stability, the temperatures at all points remain constant, which is steady conduction. The fundamental equation for conduction is

$$q = kA \frac{dt}{dL} \tag{1}$$

If heat conduction is steady, the rate of heat transfer through homogeneous flat bodies can be calculated by the equation

$$q = \frac{kA (t_1 - t_2)}{L} \tag{2}$$

where q = rate of heat transfer, Btu/h
               k = thermal conductivity Btu/(ft)(h)(F°)
               A = cross-sectional area of body taken normal to the direction
                    of heat flux, sq ft
    t_1 and t_2 = temperatures at two faces, F°
               L = thickness of body, ft

Convection is heat transfer from one part of a fluid to another by the mixing of warmer with cooler particles of the fluid. For example, consider a pipe conveying water, the pipe being heated by a flame. Heat is transferred through the pipe to the water in the pipe by conduction. The water molecules become turbulent by moving through the pipe and by becoming more buoyant as they are heated, tending to rise or mix with the main body of the water, thus transferring heat by convection. Convection caused by thermal action only on the molecules is "natural" convection. Convection caused by forcing the fluid to move or flow, as by a pump, is "forced" convection.

The film coefficient of convection h is defined as the rate of heat transfer between the retaining wall and the fluid per unit area of the retaining wall, and per degree of temperature difference between the surface of the wall and the main body of the fluid. The equation for this relation is

$$q = hA(\Delta t_1) \tag{3}$$

where q = rate of convection, Btu/h
      h = film coefficient, Btu/(sq ft)(h)(F°)
      A = area of the retaining wall, sq ft
      Δt_1 = temperature difference between the surface of the wall and the
            main body of the fluid, F°

The following factors affect the film coefficient: (1) heat transfer in which the fluid does not change phase, (2) heat transfer in which a vapor is condensed, (3) heat transfer in which a liquid is evaporated. These factors are considered in later examples.

The overall coefficient is the net result of a system of convection coefficients, for example, as follows:

Internal coefficient
External coefficient
Retaining wall resistance coefficient
Fouling factors

The overall coefficient is the reciprocal of the sum of the reciprocals of each coefficient:

$$\frac{1}{U} = \frac{1}{h_1} + \frac{1}{h_2} + \frac{L}{k} + \frac{1}{h_3} \tag{4}$$

The value of temperature difference (Δt_1) is usually the log mean temperature difference of the heat transfer system, as follows:

$$\Delta t_{Lm} = \frac{t_1 - t_2}{2.3 \log_{10}(t_1 - t_2)} \tag{5}$$

where Δt_{Lm} = log mean temperature difference, F°
     t_1 = the greater terminal temperature difference, F°
     t_2 = the smaller terminal temperature difference, F°

This equation is valid for either parallel flow or counter flow but not for cross flow and cross-counter flow, which require the application of correction factors. See the Appendix for curves stating these correction factors.

We will use appropriate curves rather than cumbersome formulas for determining convection coefficients.

Radiation is the transfer of heat from one body to another as a result of the bodies' emission and absorption of radiant energy. All matter has the property of both emitting and absorbing such energy to a greater or lesser degree, the effect on the matter being the same as if heat were absorbed from or added to it, except in rare cases where photochemical reactions are involved. Radiant energy travels through space at the velocity of light. Unlike heat, radiant energy does not require the presence of matter for its transmission. As an example, the earth receives radiant energy from the sun even though it is separated from the sun by an almost perfect vacuum.

In practice, heat transfer by more than one of the three methods is usually involved. It is, therefore, necessary to be able to recognize and calculate the rate of heat transfer by each method in order to design practical heat transfer equipment.

A practical method for determining radiation between two bodies at different temperatures is illustrated by Fig. A-j-1 in Appendix J. This is true if it is assumed that all radiation transmitted by the emitting surface is absorbed by the receiving surface; in other words, emission is 100%. Radiation by flame to surrounding walls and to radiant heating surface is a more complex mechanism. The mathematical computations are cumbersome. The "Maker" chart for radiation is a simple, graphical solution that requires the following input:

1.   Combustion heat release rate, Btu/h
2.   Combustion temperature, F°
3.   Total combustion chamber volume
4.   Total combustion space wall area
5.   Effective radiant heating surface, sq ft
6.   Radiant heating surface temperature, F°

This chart is illustrated with examples in Appendix Fig. A-i.

If blackbody radiation passes through a gas mass containing, for example, carbon dioxide, absorption occurs in certain regions of the infrared spectrum. Conversely, if the gas mass is heated, it radiates in those same wavelength regions. This infrared spectrum of gases has its origin in simultaneous changes in the energy levels of rotation and of interatomic vibration of the molecules; at the temperature levels in industrial furnaces, it is of importance only in the case of gases like carbon dioxide, carbon monoxide, the hydrocarbons, water vapor, sulfur dioxide, ammonia, and some other heteropolar gases with emission bands of sufficient magnitude to merit consideration. The gases with symmetrical molecules, hydrogen, oxygen, nitrogen, and so on do not show absorption bands in those wavelength regions of importance in radiant-heat transmission at temperatures met in industrial processes.

In most heat recovery equipment, temperatures are comparatively low and the radiant beam is short because of close tube spacing. Therefore, gas emission is low, and the overall effect on total heat transfer is not too important and can usually be ignored. However, the effect of gas radiation should not be overlooked in fired furnaces and boilers.

The various conditions of heat transfer involved in the design of heat recovery equipment of the type covered are presented by the following:

Ambient temperature and altitude correction factor (for gas turbine applications)—Fig. 3

Gas pressure drop versus mass flow and number of passes—Fig. 4

Approximate mean specific heat of fuel gas—Fig. 6

Base value of film coefficient for gases heated or cooled inside tubes, turbulent flow—Fig. 14

Diameter and temperature correction factors for Fig. 14—Fig. 15

Base value of film coefficient for gases at atmospheric pressure heated or cooled inside tubes, turbulent flow—Fig. 16

Diameter and temperature correction factors for Fig. 16—Fig. 17

Gas flow outside tubes, gas mass flow–tube diameter factor versus mass flow—Fig. 18

Temperature correction factor versus absolute temperature for Fig. 18—Fig. 19

Base value of the film coefficient gas heated outside horizontal tubes, natural convection—Fig. 20

Temperature, diameter, and pressure correction factors for Fig. 20—Fig. 21

Gas-heated inside or outside vertical plates or tubes, natural convection—Fig. 22

Temperature and pressure correction factors for Fig. 22—Fig. 23

Base value of pressure drop, gases inside tubes, turbulent flow—Fig. 25
Correction factors for diameter and temperature for Fig. 25—Fig. 26
Curve for converting 90° bends into additional straight pipe—Fig. 27
Gas, pressure drop at entrance to tubes—Fig. 28
Friction factor versus Reynolds number for air and gases—Fig. 29
Finned circular coil, gas pressure drop versus gas mass flow—Fig. 38
Gas pressure drop through portions of heat recovery unit—Fig. 39
Base value of film coefficient for liquids inside tubes, turbulent flow—Fig. 40
Temperature and tube diameter correction factor for Fig. 40—Fig. 41
Fin efficiency chart—Fig. 42
Water pressure drop in coiled pipes—Fig. 43
Loss of head (in ft) of any liquid from sudden enlargement—Fig. 44
Loss of head (in ft) of any liquid from sudden contraction—Fig. 45
Heat transfer coefficients, Monsanto Therminol 44—Fig. 47
Heat transfer coefficients, Monsanto Therminol 55—Fig. 48
Heat transfer coefficients, Monsanto Therminol 60—Fig. 49
Heat transfer coefficients, Monsanto Therminol 66—Fig. 50
Heat transfer coefficients, Union Carbide Ucon HTF–14—Fig. 51
Heat transfer correction factors for Fig. 51—Fig. 52
Heat transfer coefficients, Dow Chemical Dowtherm "G"—Fig. 53
Friction factor versus Reynolds number for liquids—Fig. 54
Pressure drop curves, Monsanto Therminol 44—Fig. 55
Pressure drop curves, Monsanto Therminol 55—Fig. 56
Pressure drop curves, Monsanto Therminol 60—Fig. 57
Pressure drop curves, Monsanto Therminol 66—Fig. 58
Pressure drop curves, Dow Chemical Dowtherm "G"—Fig. 59

PROPERTIES OF EXHAUST GAS

Exhaust gas from different sources varies in character. Obviously, it is important to understand these characteristics to evaluate properly a particular heat recovery problem. For this reason, Chapter 1 is devoted to the subject of gas properties before treating the several heat recovery systems described in the introduction.

AIR

Since most exhaust gas is derived from air or consists in great part of air, and since that air itself enters into many heat recovery processes, we will deal with air first.

Air is a mechanical mixture of the gases oxygen (O_2) and nitrogen (N_2), with about 1% of argon (A). Moreover, atmospheric air of ordinary purity also contains about 0.04% of carbon dioxide (CO_2). The average percentage composition of air is as follows:

	N_2	O_2	A
By volume	78.06	21.0	0.94
By weight	75.50	23.2	1.30

The weight of pure air at 32°F and a barometric pressure of 14.696 lb per sq in. is 0.08071 lb per cu ft. The weight of so-called standard air at 69°F sea level is 0.075 lb per cu ft. This is for "dry" air containing zero water vapor. The density (weight per cu ft) of air at other temperatures and pressures is expressed by the formula

$$\rho = 0.075 \frac{(529)}{(T_1)} \frac{(P_1)}{(14.7)} \tag{6}$$

where ρ = air density, lb per cu ft
    T_1 = air temperature, R° absolute
    P_1 = air pressure, psia

Figure 1 illustrates the weight of "dry" air at various temperatures, at 14.7 psia.

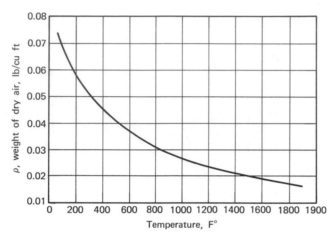

Fig. 1. Curve, weight of dry air at various temperatures.

If it is certain that the air being handled is in fact "dry" or reasonably dry, its enthalpy and other thermodynamic properties may be taken from *Thermodynamic Properties of Air* by Keenan and Kay, John Wiley & Sons, New York, 1945.

The presence of significant quantities of water vapor in air has a marked effect on specific heat. The specific heat of air rises with temperature increases and with increases in water vapor content. This is illustrated by Fig. 2 for various water vapor contents by weight and for various absolute temperatures.

GAS TURBINE EXHAUST

The exhaust from gas turbines is composed of air and combustion products. Since gas turbines operate with large amounts of excess air, most of the exhaust content is air whether the turbine is fueled with liquid fuel or natural gas. The excess air is used to cool the combustion chamber and reduce the temperature of the combustion gases to a point that can be tolerated by the turbine parts. The expansion of the gases through the turbine stages results

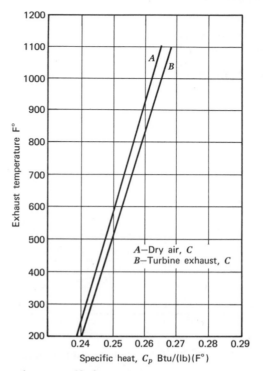

Fig. 2. Gas turbine exhaust specific heat versus temperature.

in an average of 18% oxygen-laden turbine exhaust at temperatures of 725 to 1100°F. This exhaust gas can be used to generate additional useful heat or energy. Gas turbines with recuperators have much lower exhaust gas temperatures, in some cases as low as 525°F. Obviously, the reduced exhaust gas temperature results in higher gas turbine efficiency and consequently less heat to be recovered.

The addition of combustion products, particularly the water vapor content, increases the average specific heat of the gas in Fig. 2. Curve *A* is for pure air; Curve *B* is for the typical products of gas turbine exhaust.

Altitude must also be taken into consideration, since altitude affects turbine mass flow. Altitude correction factors for various ambient temperatures are shown by Fig. 3.

Ambient temperature also affects mass flow and exhaust temperature. This effect must be considered in consulting the gas turbine performance curves. It is essential that the performance curve for its respective gas turbine be available, particularly if part-load heat recovery must be determined.

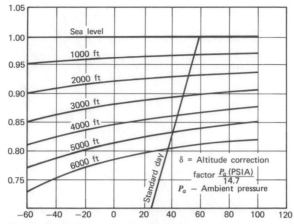

Fig. 3. Altitude correction factors.

Gas turbine exhaust is usually relatively clean, especially exhaust from turbines fueled by natural gas. Light-oil-fired turbines have some residue in their exhaust, particularly during radical load changes. Generally, the use of extended surface (finned) heat recovery tubes is recommended. However, some form of soot blowing and/or cleaning means should be included in heat recovery equipment used with liquid-fueled gas turbines.

The allowable gas pressure drop for gas turbine heat recovery equipment can vary from 3 in. water column to 6 in. maximum. Figure 4 illustrates approximate gas pressure drop versus mass flow at various average temperatures and at numbers of gas passes.

RECIPROCATING ENGINE EXHAUST

The characteristics of exhaust gas from reciprocating engines depends on whether the engine is oil or gas fired. The firing mode must be determined before heat recovery can be properly evaluated. Oil-fueled engines generally have a "dirtier" exhaust than engines fueled by natural gas, particularly if the fuel-air ratio is out of adjustment or at rapid load changes. The water vapor content of natural-gas-fueled exhaust is higher than that of oil-fueled engines. The result is a higher average specific heat, as illustrated by Fig. 5.

The exhaust gas temperature varies from approximately 800 to 1350°F, depending on the size of the engine, its efficiency, and whether it is supercharged.

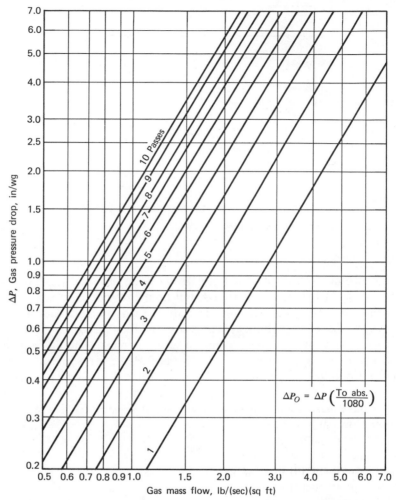

Fig. 4. Gas pressure drop versus mass flow for number of passes and average gas temperature 620°F.

Light-oil combustion yields approximately 8.6% water vapor in its exhaust; natural gas combustion yields approximately 18.2%.

We do not deal here with exhaust gas analysis but only with water vapor content, which has a marked effect on specific heat, as previously stated. The other constituents are CO_2 and H_2, with some SO_2 in small quantities for light oil.

Sulfur-bearing oil must be used advisedly since, together with water vapor, it causes corrosion and thus limits the choice of heating surface materials.

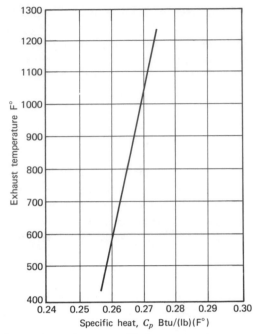

Fig. 5. Reciprocating engine exhaust specific heat versus temperature F°.

INCINERATOR EXHAUST

Incinerator exhaust gas analysis can vary considerably, depending on the material incinerated. This must be determined and exhaust gas characteristics known. Several common types of incinerators lend themselves to heat recovery: (1) continuously fed municipal solid waste, (2) sewage sludge, (3) waste liquid, and (4) waste gas incineration.

Batch-type incinerators such as those used by apartment, buildings, schools, and the like do not warrant heat recovery except perhaps air-to-air type for the larger models. Incinerators used in hospitals, particularly in large medical centers, are usually sufficiently continuous, with a feed rate warranting some type of heat recovery, possibly in the form of process steam; this steam can then supplement the hospital's steam system. Hospital applications, however, do cause problems, namely, the organic material incinerated and the incineration of plastics that produce corrosive gas from the formation of HCl. Under such conditions, the heat recovery heating surface must be constructed of corrosion-resistant material. Refer to the Appendix for a list of materials resistant to various corrosives.

Waste disposal by incineration, with effective heat recovery from the incinerator stacks' gases, is a profitable solution to the problem posed by the increase in more uniformly restrictive air and water pollution regulations. Economical incineration for many businesses and institutions will permit them to continue their operations as landfill dumps are phased out and the open burning of refuse is banned in a large part of the United States.

For businesses and institutions already equipped with or contemplating a total energy plant, however, the increasing cost of waste disposal can be converted to a profit by adding a heat recovery incinerator to the system to recover substantial amounts of heat from the refuse.

This concept is particularly applicable to shopping centers, office buildings, schools, and large apartment complexes, where refuse with high heat value is normally encountered, and where incinerator operations can be timed to supply steam during hours of reduced load on the engine generators. The availability of such low-cost by-product steam during off-peak hours is invariably a pronounced economic advantage for absorption air conditioning.

Municipal Solid Waste

The use of refuse incinerators as a heat source is by no means new. This technique has been relatively common for years in Europe, where entire communities are fueled in that manner by central plants. It is ironic that only the pressure of increasingly stringent air quality regulations has triggered widespread use in the United States.

Each proposed installation must, of course, be surveyed on an individual basis to determine the amounts, types, and heating value of the wastes involved. For preliminary study purposes, however, the following classifications are useful (heating values are approximate):

Type 1—A mixture of highly combustible material consisting of paper, cardboard cartons, wood boxes, and the like, with a maximum 10% by weight of plastic bags, coated paper, laminate paper, treated corrugated cardboard, oily rags, and plastic or rubber scraps. The mixture typically contains 10% moisture and has an "as-fired" heating value of 8500 Btu/lb.

Type 2—A mixture consisting of approximately even portions of Type 1 materials and garbage by weight. It may contain as much as 50% moisture and have an as-fired heating value of 4500 Btu/lb.

Type 3—Garbage consisting of animal and vegetable wastes from food service institutions, hotels, supermarkets, hospitals, and the like, typically with as much as 70% moisture and an as-fired heating value of 2500 Btu/lb.

Sewage Sludge

Type 4—Sewage sludge, consisting of municipal sewage with quite varied contents. These include human waste, metal and glass objects, rags and paper, greases from meat packing plants, and so on. The solids content of sewage sludge varies from 5 to 7%, with 5% the more usual amount. We do not dwell on this point but state that the final, treated, dewatered sludge "cake" usually contains about 35 to 45% solids.

The following list is a typical sludge cake condition:

Feed cake to furnace	15,500 lb/h
Moisture	10,075 lb/h
Solids	5,425 lb/h
Inerts	2,170 lb/h
Volatiles	3,255 lb/h

Sludge composition is as follows (decimal fraction by weight):

$C = 0.558$
$O_2 = 0.310$
$H_2 = 0.082$
$N_2 = 0.050$
$S = 0.000$

The average heating value is 10,000 Btu/lb, which is, of course, affected by the volatiles content of sludge and the moisture content, which must be determined for each case by actual analysis.

A sludge such as that just described at the given rate produces the following incinerator off-gas conditions, based on 50% excess air and afterburning to eliminate all traces of unburned hydrocarbons:

Gas temperature—1200°F
Dry gas total—40,258 lb/h
Water vapor total—12,477 lb/h
Mixture total—52,736 lb/h

The exhaust gas will contain

$CO_2 = 6,660$ lb/h
$N_2 = 30,426$ lb/h
$SO_2 = $ zero lb/h
$O_2 = 3,172$ lb/h

This is the condition that the heat recovery equipment will receive as input gas. The resulting heat exchange is discussed later.

Note that there are no corrosive products in this particular gas, although O_2 produces oxygen corrosion if final stack temperatures approach or are less than the dew point, which is never the case in this kind of heat recovery. Furnace off-gas temperatures can vary from 1000 to 1400°F. It is generally not considered good practice to allow the temperature to go below 1000°F because of the possibility of carryover of unburned hydrocarbons.

A problem to be considered in the design of heat recovery equipment for any kind of solid waste incinerator is ash carryover. Most of the ash is discharged from the ash outlet of the furnace, but some is carried over with the gas, usually 10 μ or less. The amount of ash is never constant; therefore, the heat recovery equipment must be designed with this in mind. Ash hoppers and suitable soot blowers must be provided.

Typical quantities of solid waste material (exclusive of sewage) may be estimated as follows:

TABLE 1   Estimates of Solid Waste Production

Installation	Daily Quantity	Type
Apartments	4 lb per person	1, 2
Cafeterias	$\frac{1}{2}$ to $\frac{3}{4}$ lb per meal	1, 2, 3
Churches	10 lb/100 seats	1, 2, 3
Department stores	1 lb/30 ft of floor area, food service $\frac{1}{2}$ to $\frac{3}{4}$ lb per meal served	1 3
Hospitals	7 lb per bed	1, 2, 3
Hotels, motels	2 lb per room plus 2 lb per meal served	1, 2, 3
Office buildings	1 lb per 100 sq ft of floor area	1
Restaurants	2 lb per meal served	1, 2, 3
Schools	$\frac{1}{4}$ lb per student $\frac{1}{2}$ to $\frac{3}{4}$ lb per meal	1, 2
Shopping centers	varies with type and locale	
Supermarkets	1 lb per 25 sq ft of sales area per operating hour	1, 2, 3
Municipal sewage	about 200 gal/day per person	4

Auxiliary fuel requirements are as follows, Btu/(h)(lb):

Type 1   1000
Type 2   1500
Type 3   3000
Type 4   zero to 850

Industrial wastes differ greatly from "standard" municipal solid wastes in chemical content and heating value. The following are some components of industrial refuse:

TABLE 2 Heating Values of Some Wastes

Material	Ash	H_2O	Cl	S	Low Heating Value (Btu/lb)
Car tires	6.3%			1.2%	15,570
Used fats	fluctuates			0.7	13,700
Refinery residues	3.5	20–30%		0.7–1.5	9,880–10,800
Oil sludge	40.5	30		0.5–1.5	3,860
Oil carbon	27.0			4.5	11,700
Tank residues	1.0	30–50		0.5–1.5	7,190–9,880
Tar residues	3.0			1.1	13,500
Acid sludge	fluctuates			12–20	7,190
Oil filter residues	53	3.5		0.6	6,920
PVC residues	0.5		48.5%		8,080
Polyethylene residues	traces				18,000

Figure 6 illustrates the relations among temperature, moisture content, and specific heat for combustion products.

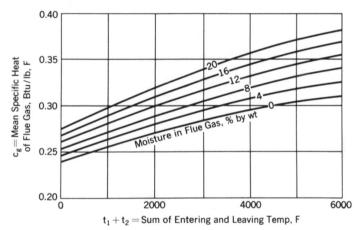

Fig. 6. Specific heat of water-vapor-laden air versus temperature. Courtesy Babcock & Wilcox.

In general, regardless of the source of the exhaust gas used in heat recovery, the following characteristics must be known:

Total gas flow—lb/h
Water vapor flow—lb/h
Dry gas flow—lb/h

Gas temperature—F°
Average specific heat
Chemical content, particularly corrosives
Allowable gas pressure drop
Type of heat recovery
   gas-air
   gas-water
   gas-water-steam
   gas-organic heat transfer fluid
Ash content of exhaust gas
Stack temperature
Heat recovery, Btu/h

Before a heat recovery evaluation can be made, the preceding list of factors must be considered and the characteristics of the gas in question must be determined. These subjects are dealt with in the following chapters. Equipment form and design is selected, with accompanying calculations provided.

GAS-AIR HEAT RECOVERY

USES AND APPLICATIONS

Gas-to-air heat exchange is used most commonly as a recuperator to heat combustion air by heat exchange with flue gas. It is also used in direct-fired air or gas heaters used in process work. One such application is the Escher-Wys closed-cycle gas turbine system, where nitrogen gas is heated in a fired heater to the temperature required by the cycle.

Here we discuss only indirect fired equipment; such equipment has a barrier between the combustion products and the air or gas being heated.

This mode of heat recovery is used when there is no application for gas to liquid heat exchange or steam generation. Very often, however, an air heater or recuperator is used on high pressure boilers after the economizer to heat combustion air prior to its introduction into the combustion chamber. Recuperators are also used in gas turbine cycles where the turbine exhaust heats air after the compression stages before its entrance into the combustion chamber.

ECONOMICS OF GAS-AIR HEAT RECOVERY

Generally, an air-to-air or gas recuperator is economical, providing that it is properly designed and used to optimum efficiency. Figure 7 illustrates this point. The curves show payback in dollars per hour for various air flow rates and values of temperature change. For example, if we have an air flow rate of 24,000 lb per hour and the temperature change is 900°F, the payback rate is approximately $3.07 per operating hour based on fuel gas cost of 60 cents per 1000 cu ft. Payback for other gas rates can be computed by direct ratio. This payback value then has to be compared to the capital investment, amor-

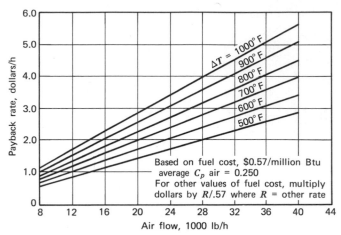

Fig. 7. Air heater payback versus flow and ΔT.

tization, interest, maintenance, insurance, operation, and other costs to determine the payoff period. An example follows:

Air heater cost	$30,000
Installation cost	18,000
Total cost	$48,000

Target time for amortization, 3 years.
Interest rate, 9% per year, declining balance.

Write-off of cost, 3 years	$16,000 per year
Interest average $\dfrac{(3 + 1)}{(3)} \dfrac{(0.09)(48,000)}{(2)}$	2,880 per year
Maintenance estimate	3,600 per year
Insurance, 1% of capital cost	480 per year
Taxes, $1\frac{1}{2}$% of capital cost	720 per year
Total owning and operating cost	$23,680 per year

Annual fuel saving based on fuel gas cost of 60 cents per 1000 cu ft at $3.07 per hour, continuous operation, is $26,893. At this rate, the facility could be shut down 1047 hours per year and still pay back in three years. After the three-year period, the write-off and interest costs are a net gain.

MATERIALS AND TEMPERATURE LIMITATIONS

Materials of construction for air heat exchangers depend on operating pressure and temperature. In most cases, operating pressure is at or near atmospheric such as in recuperators for boilers or incinerators. In these cases, we can use either plates or tubes. Air heaters operating at significantly high pressures generally require tubes. In addition to temperature, corrosive conditions must be considered in the selection of material.

Table 3 gives maximum metal temperatures for continuous operation for various materials.

TABLE 3   Maximum Metal Temperatures

Temperature (°F)	Material
400	copper
550	brass
700	copper-nickel
750	carbon steel (ASME Spec. SA53)
850	carbon steel (ASME Spec. SA178)
900	low alloy steel (ASME Spec. SA209)
1000	low alloy steel (ASME Spec. SA217)
1450	high alloy steel (ASME Spec. SA213, SA304, SA312, SA240, etc.)
1500	high alloy steel (19–9)
1800	inconel-X, alloy 25
1900	hasteloy-C

Allowable stresses for the condition in question should be taken from material handbooks or the ASME Code.

The greater the temperature change, the more efficient the process becomes. Therefore, the air or gas stream should be heated to the greatest possible extent. There are limitations, however. For a boiler recuperator, where heated combustion air enters the combustion chamber around the burners, heated air temperature rise should be limited to about 1000°F. On the cold end, the final exit temperature should be well above the dew point temperature for the condition. Sulfur content of the flue gas raises allowable minimum metal temperature because the dew point temperature is reduced. Refer to Fig. 8, which illustrates minimum metal temperature versus sulfur content for several fuels.

CORROSION AND PREVENTION

Corrosion in the tubes low temperature zone is of primary concern in the design and operation of air heaters. Experience indicates that the best preventive is an air heater designed to avoid dangerously low tube temperatures at every stage throughout the entire operation so that tube temperature is

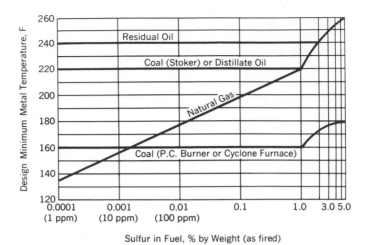

Fig. 8. Minimum metal temperature versus sulfur content. Courtesy Babcock & Wilcox.

always substantially higher than the dew point temperature. The tube metal temperature at any point in an air heater is approximately equal to the average of the adjacent gas and air temperatures, since the gas and air heat transfer rates are about equal. In a counterflow air heater the lowest tube metal temperatures occur in the air inlet and exit gas zone, where both the air and gas are at their lowest temperatures. When operating with inlet air of 100°F and uptake gas temperatures of 300 to 350°F, the tube metal temperature in the coolest zone will be about 200 to 225°F. Under normal conditions, when firing fuel oil of low sulfur content, the dew point of the combustion products will be about 50°F below the tube metal temperature in the coolest zone. Air heater corrosion can be reduced or practically eliminated in several ways.

The air heater tube metal temperature can be increased by designing the boiler for higher exit gas temperatures. However, boiler efficiency will be reduced.

By using parallel flow instead of counterflow arrangements, the inlet air enters the air heater in a hotter area. The tube metal temperatures will increase but the temperature difference between air and gas will be lower, thus reducing the heat absorption in the air heater and consequently the boiler efficiency.

Use of air bypass dampers for tube temperature control provides a good reserve margin against corrosion, with loss of boiler efficiency only at light load. By operating the dampers so that part of the combustion air bypasses the air heater during lighting-off and low-load operation, the tube temperatures will approach the surrounding gas temperatures.

The tube metal temperatures may be increased by recirculating a part of the exit (hot) air through the entering air pass to raise the inlet air temperature. The loss in efficiency is negligible, but some increase in fan power is required.

To reduce or eliminate tube corrosion, some consideration has been given to the use of alloy steel or aluminum, but possible advantages are usually offset by the additional cost. Steam air heaters are also sometimes used to increase tube temperature.

The life of the air heater can be lengthened considerably by the use of thicker wall tubes, if the conditions affecting corrosion cannot be corrected.

AIR HEATER FIRES

If soot is allowed to accumulate on air heater surfaces, there is some danger of fire. Experience proves, however, that soot fires in air heaters are due to malfunctioning. No such fires will occur if proper operating procedures are followed. Investigations indicate that air heater soot fires have occurred almost without exception while lighting off, operating at low rating, bringing up the rating after extended periods of low-fire operation, or in cases of intermittent firing of the burners with insufficient combustion air.

CLASSIFICATION OF AIR HEATERS

Air heaters may be classified according to their operation principle as (1) recuperative and (2) regenerative. Among successful air heaters operating on the recuperative principle are tubular (Fig. 9) and plate types, with the heat provided by (1) flue gases of the boiler, (2) steam in coils, or (3) separately fired furnaces. Among air heaters operating on the regenerative principle are the rotary regenerative (Fig. 10) (Ljungstrom) and the diphenyl (Fig. 11), with heat provided by flue gases, and the pebble type (Fig. 12), with heat provided by flue gases or a separate furnace.

In the recuperative design, heat is transferred directly from the hot gases or steam on one side of the surface to air on the other side.

In a regenerative heater, heat is transferred indirectly from the hot gases to the air through some intermediate heat conducting medium, as in rotary regenerative, diphenyl, and pebble heaters. These heaters are used in applications other than steam generation, such as open-hearth furnaces.

TUBULAR AIR HEATER

The tubular air heater is essentially a nest of straight tubes expanded into tube sheets and enclosed in a suitably reinforced steel casing. The casing

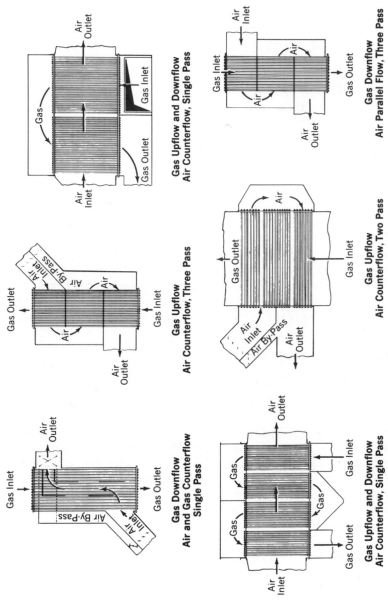

Fig. 9. Outline diagram of several tubular air heaters. Courtesy Babcock & Wilcox.

functions as the air or gas flue and is provided with air and gas inlet and outlet openings and the necessary baffles, hoppers, and supporting members. The arrangement may be either vertical or horizontal. In the modern tubular air heater, the tubes are rolled into tube sheets at both ends. To provide for expansion, one tube sheet should be free to move with respect to the casing without air leakage. Tube sizes may range from 2 to 4 in. Several factors are involved in establishing tube size. With smaller tubes, the surface is used more effectively, but the tubes are more difficult to clean. Greater space is taken by the large-tube heater. The overall cost of the heater—including the manufacture, shipment, and erection of the tubes, casing, and supports—and building cost tend to be minimal for some particular size of tube. For stationary boilers, 2-in. and $2\frac{1}{2}$-in. tubes are generally used; in marine practice, $1\frac{1}{2}$-in. tubes are used. Figure 9 outlines some common flow arrangements for tubular heaters.

ROTARY REGENERATIVE AIR HEATERS

The rotary regenerative air heater generally consists of multiples of slightly separated metal plates supported in a frame attached to a slowly moving rotor shaft, which is arranged edge-on to the gas and air flow. As these plates pass progressively through the gas stream, they are heated and then, in passing through the air stream, they give up heat to the air before again entering the hot gas stream, thus maintaining the regenerative cycle. The plates are placed in segmental groupings to complete a circle around the shaft. Seals are provided to reduce air infiltration into the gas. Soot blowers are located in the assembly to remove soot and oil deposits periodically. See Fig. 10 for a diagrammatic illustration of a rotary air heater.

PLATE-TYPE AIR HEATER

The plate-type air heater generally is made of thin, flat, parallel plates with alternate wide and narrow spaces to match the ratio of gas weight to air weight. Thus, the flue gas is forced through the narrow spaces with countercurrent flow. To obtain more heating surface for a given volume, the gas space is made as small as possible within the limits required for cleaning. Commercial designs have differed chiefly in the method of sealing the air spaces from the gas spaces. Many patented constructions have been tried, but only the welded joint has survived.

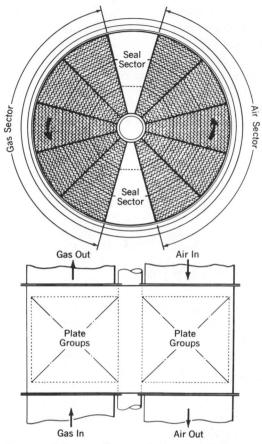

Fig. 10. Outline of rotary air heater. Courtesy Babcock & Wilcox.

STEAM HEATED AIR HEATERS

Steam heated air heaters are sometimes used to reduce fouling and corrosion by increasing the minimum tube metal temperatures by heating the combustion air before it enters the air heater.

SEPARATELY FIRED AIR HEATERS

Separately fired air heaters are used when hot air is required for an industrial process. Air at atmospheric or higher pressures can be heated to the desired temperature by using a separately fired air heater. Such a heater usually consists of a refractory furnace with a tubular heater arranged for a number of air and gas passes. To prevent overheating of the tubes, it is essential to control the gas temperature entering the heater. This may be done by (1) diluting the gas with excess air or (2) recirculating the gas from the air heater outlet to the furnace, thus reducing the temperature of the furnace gases. The latter method is the more efficient.

DIPHENYL OXIDE AIR HEATERS

Occasionally, an air heater operating on the regenerative principle, can be applied where the intermediate heat conducting mechanism is diphenyl oxide or a similar suitable liquid. The liquid is heated as it passes through a heat exchanger located in the flue gas stream and cooled as it passes through another heat exchanger located in the air stream. By continuous recirculation of the liquid, heat is transferred from the gas to the air. See Fig. 11 for a schematic diagram of such a system.

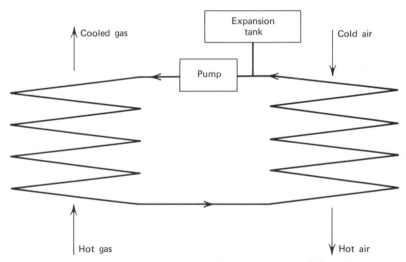

Fig. 11. Schematic of liquid-air regenerative system using diphenyl.

PEBBLE AIR HEATERS

Pebble heaters are designed to heat air and other gases to higher temperatures than is economically feasible using the recuperative principle. Arrangement of a pebble heater, designed for heating air from ambient temperature to over 2000°F, consists of a vertical refractory lined column divided into two sections. The upper section is filled with pebbles that are heated by burners to the required temperature. The pebbles drop into the lower section, through which cold air is circulated in counterflow relation to the flow of hot pebbles. The cooled pebbles from the lower portion of the lower chamber are then conveyed to the top of the heater section for reheating. Various materials may be used for the pebble heat transfer medium. Spherical pebbles $\frac{1}{2}$ in. in diameter of mullite composition (72% Al_2O_3, 28% SiO_2 are commonly used as the heat conducting medium. In addition to heating air or steam, the pebble heater has other applications, among which are heating hydrogen or natural gas for the reduction of metallic oxides and producing ethylene from a hydrocarbon feed source. See Fig. 12 for a diagram showing a pebble air or gas heater.

FOULING OF AIR HEATERS

When dirty air or gas is used in the air heater, such as from an incinerator, fouling must be prevented. Dust and cinder disposal hoppers are required under the tubes to suit flue conditions. Soot blowers are located so that a jet of steam, air, or water can be blown through each tube as required or across the tubes when gas crosses the tubes on the outside. Ample flow paths must be provided for the gases, and where space permits, dust hoppers can be fitted ahead of the air heater. The result is a significant reduction in the amount of dust and flyash entering the heater. Instead of a hopper at the heater inlet, a high velocity cyclone can be used to remove the bulk of ash carryover from the furnace. When a cyclone is used, however, an induced draft fan must be installed after the air heater.

In marine practice, the use of air heaters and regenerative feed heating results in high plant efficiency and overall economy. Since the air heater, unlike the economizer, is not a pressure vessel, its design can be relatively simple. An important additional advantage is the improvement in furnace combustion due to preheated air, particularly at both the low and high rates of firing. This will reduce soot accumulation and lessen the likelihood of ignition loss.

AIR HEATERS FOR MARINE SERVICE

For marine service, air heaters are nearly always of the tubular type; the plate and regenerative types are seldom used. Tubular air heaters in marine service are generally of the horizontal type. Since vertical units require too much space, they are seldom used. In the horizontal type (to avoid lodgement

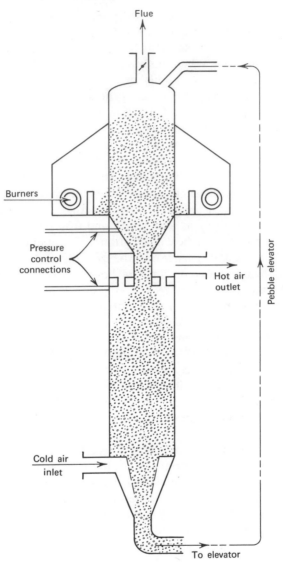

Fig. 12. Pebble heater diagram. Schematic arrangement of pebble air heater.

of deposits inside the tubes), the air to be heated passes through the inside of the tubes and the hot gases pass over the outside of the tubes, while the reverse is true of the vertical type.

The tube arrangement is usually "in-line," which makes external cleaning easier, more than offsetting the slight advantage in heat transfer with staggared tubes. Obviously, increased air pressure must be provided when an air heater is installed to overcome the additional resistance to air flow through the air heater, air ducts, and burner. For boilers of the same size and rating, a higher total air pressure will usually be needed for an air heater installation than for an economizer. Where both an air heater and an economizer are used, required air pressure is still higher.

Air heater tubes for marine work are generally $1\frac{1}{2}$ to $2\frac{1}{2}$-in. o.d. The $1\frac{1}{2}$-in. tube is used in most instances. Tube diameters must be selected with due consideration for both draft loss and heat absorption. To reduce internal flow resistance, which varies inversely as the i.d., the larger tubes should be used. On the other hand, to provide the greatest surface in the limited space available, small tubes should be used. Furthermore, at any given gas mass flow, the heat transfer rate will increase as the tube diameter is reduced, since the heat transfer rates of the gases flowing across and through the tubes, respectively, vary inversely as about 0.53 and 0.34 powers of the tube diameter. To obtain maximum heat absorption with minimum fan horsepower, the effect of tube size and spacing must be carefully analyzed.

When an air heater is used on a marine boiler, the temperature of the inlet air to the air heater ranges from 80 to 100°F, and at normal operating rates, the exit air temperatures range from 300 to 350°F. Any variation in the inlet temperature affects air pressure, efficiency, and external tube corrosion, and therefore must be considered in the design. Exit gas temperatures leaving the air heater are generally reduced to 300 to 350°F required for boiler efficiencies of 88 to 87%, respectively. Because of the lesser weight and the lower specific heat of the air, as compared with the combustion products, the reduction in temperature of the hot gases passing through the heater when using 15% excess air and firing oil is about 13% less than the difference between the exit and inlet air temperatures.

For the same heat absorption, since the heat transfer coefficients of the gas and air films on their respective sides of the tube are of both the same magnitude and relatively quite low, air heaters require more heating surface than do boilers, economizers, and superheaters. In an air heater tube, heat must traverse an air as well as a gas film; while in a boiler, economizer, or superheater tube, the heat, after traversing the gas film, traverses a water or steam film that has a much higher transfer coefficient than the air film of the air heater.

RECUPERATORS FOR GAS TURBINES

Gas-to-air recuperators (or regenerators) are now also used on marine, industrial, and utility open-cycle gas turbine applications. The gas turbine recuperator receives air from the turbine compressor at pressures ranging from 73.5 to 117 psia and temperatures from 350 to 450°F. Gas turbine exhaust gas passes over the other side of the recuperator at exhaust temperatures ranging from 750 to 1000°F, depending on the turbine. The air side (high pressure side) of the recuperator is in the system between the compressor and the combustor. Compressor air is raised to a higher temperature, up to about 750 to 900°F as it enters the combustor. Turbine exhaust gas is reduced to between 500 and 650°F, from 850 to 950°F or higher. This heat recovery contributes appreciably to the turbine fuel rate reduction and increase in efficiency. However, shaft horsepower is reduced because of pressure drop through the recuperator.

Obviously, pressure drop through the regenerator or recuperator is important and should be kept as low as practical on both sides. Generally, the air pressure drop on the high pressure side should be held below 2% of the compressor total discharge pressure. The gas pressure drop on the exhaust side (hot side) should be held below 4 in. of water.

Gas turbine regenerators are usually constructed either as shell-and-tube type heat exchangers using very small diameter tubes, with the high pressure air inside the tubes and low pressure exhaust gas in multiple passes outside the tubes. Aircraft-type regenerators are also sometimes made of cores or matrixes of lightweight sheet metal formed to provide air and gas passages with the sheets furnace-brazed together.

Even with a regenerator and the resultant reduced exhaust gas temperature, further heat recovery can still be effected by the use of a heat recovery steam generator or liquid heater. For example, with a full load exhaust gas temperature of 650°F out of the regenerator, steam can be generated at 100 psig with a final stack temperature of 395 to 405°F. Such a combination results in extremely attractive fuel rates and efficiencies.

HEAT TRANSFER, GAS-AIR

Gases Inside Tubes

Internal turbulence greatly improves heat transfer at the expense of increased pressure drop. Inserting cores, retarders, fins, or other turbulence promoters into tubes increases both the internal heat transfer coefficient and the pressure drop. One of the best simple retarders is a coil of wire fitting tightly against the inner surface of the tube. This produces the greatest advantage at the lower Reynolds numbers, where, for the same heat transfer, the retarder requires only one-fourth the power loss than for the empty tube. Internally ribbed tubes with swirlers are also available. Figure 13 illustrates an internally ribbed tube. This construction is particularly advantageous if the pressure of the gas inside the tube is lower than that of the gas outside, or if there is

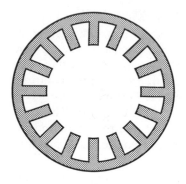

Fig. 13. Internally finned tube section.

gas inside and liquid outside. However, the use of any turbulence promotion devices increases the possibility of fouling. Therefore, only clean gas or air should be used inside such a tube.

The recommended equation for the flow of fluids inside of tubes is

$$\frac{hD}{k} = 0.0255 \left(\frac{DV\rho}{\mu}\right)^{.08} \left(\frac{Cp}{k}\right)^{0.4} \tag{7}$$

For gases, this equation is valid for either heating or cooling, provided that the value of $G = Vp$ is greater than $1650\,\rho^{0.645}$. Values of $G_{min.} = (Vp)_{min.} = 1650\,\rho^{0.645}$ are given in Table 3.

TABLE 3   Minimum Permissible Values of $G_1 = V_{1\rho}$[a]

Pressure, psig	G_1 min. $= (V_1\rho)$ min.
0	0.46
50	1.19
100	1.73
150	2.18
200	2.59
250	2.97
300	3.32

[a]G_1 = lb per sec (sq ft)
V_1 = velocity, ft per sec
ρ = density, lb per cu ft

Rather than use cumbersome equations for the solution of problems, we will use charts and graphs as much as possible. Equation 7 is reduced to curve form in Fig. 14 for several gases commonly found in air-gas heat ex-

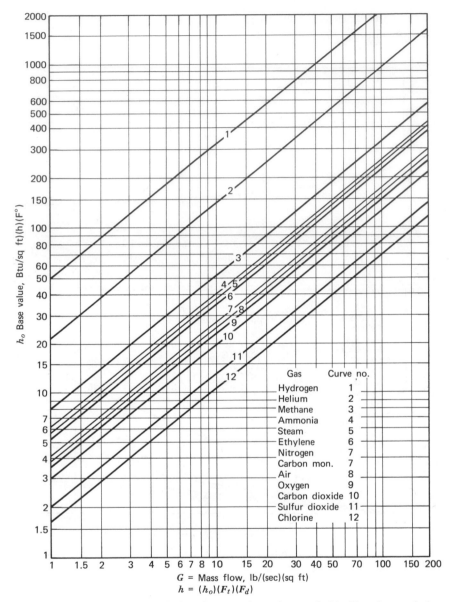

Fig. 14. Base value of film coefficient for gases heated or cooled inside tubes, turbulent flow. From Applied Heat Transmission, Stoever.

change. Correction factor curves for mean gas temperature and tube diameter are shown in Fig. 15.

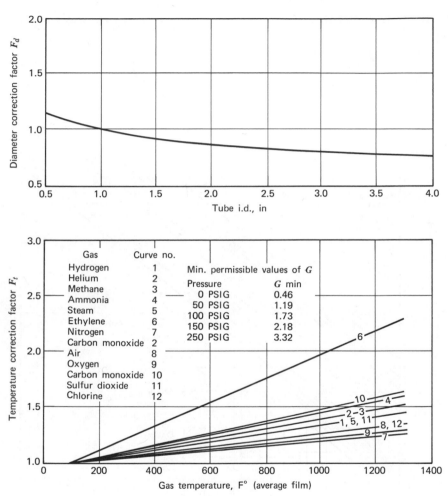

Fig. 15. Correction factors F_t for temperature and F_d for tube diameter for use with Fig. 14. From Applied Heat Transmission, Stoever.

The coefficient of convection by a gas flowing inside a tube is expressed by the equation

$$h = (h_0)(F_t)(F_d)$$

where h = film coefficient, Btu/(h)(sq ft)(F°)
h_0 = base value from Fig. 14

F_t = temperature correction factor from Fig. 15
F_d = diameter correction factor from Fig. 15

Gases at atmospheric pressure heated or cooled inside horizontal or vertical tubes, turbulent flow, are also based on Equation 7. The equation is valid for gases if V_p is greater than $1650_p^{0.645}$. The gases are assumed to obey the perfect gas law $Pv = RT$. See Fig. 16 for the base value of heat transfer

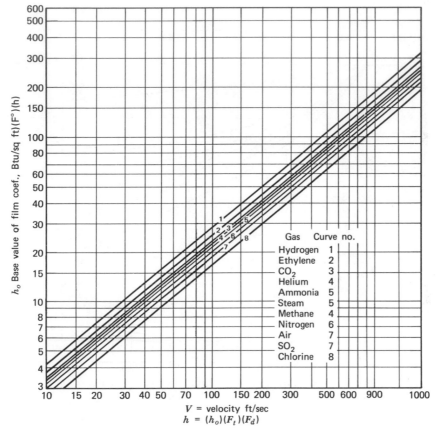

Gas	Curve no.
Hydrogen	1
Ethylene	2
CO_2	3
Helium	4
Ammonia	5
Steam	5
Methane	4
Nitrogen	6
Air	7
SO_2	7
Chlorine	8

V = velocity ft/sec
$h = (h_o)(F_t)(F_d)$

Fig. 16. Base value of film coefficient for gases at atmospheric pressure heated or cooled inside tubes, turbulent flow. From Applied Heat Transmission, Stoever.

for gases at atmospheric pressure, and Fig. 17 for temperature and tube diameter correction factors. Minimum values for V for Fig. 16 are shown in Table 4.

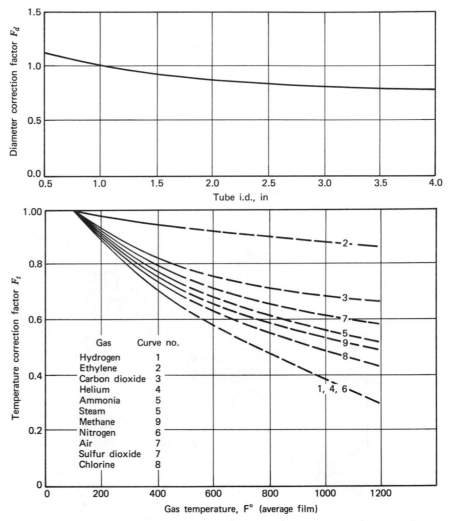

Fig. 17. Temperature correction factor F_t tube diameter correction factor F_d for use with Fig. 16. From Applied Heat Transmission, Stoever.

The coefficient of convection by a gas flowing inside a tube at atmospheric pressure, turbulent flow, is expressed by the equation

$$h = (h_0)(F_t)(F_d)$$

where h = film coefficient, Btu/(h)(sq ft)(F°)
h_0 = base value from Fig. 16
F_t = temperature correction factor from Fig. 17
F_d = diameter correction factor from Fig. 17

TABLE 4   Average Temperature of Gas, F°, V min

Gas	0	100	200	300	400	500
			Minimum Velocity			
Ammonia	9	11	13	15	17	19
Helium	39	47	55	64	72	81
Hydrogen	75	92	108	125	141	157
Methane	10	12	14	16	18	20

Gases Outside Banks of Tubes

The case of forced convection outside banks of tubes is of considerable importance, especially in air heaters and heat exchangers and in gas-to-liquid or steam heat recovery systems.

Considerable data are available for the flow of air at right angles or normal to banks of tubes. Many of these data are highly theoretical and sometimes do not bear up in practice when comparing design method with test results. The familiar equation for gases or fluids heated or cooled outside single tubes, with direction to flow normal to the tubes, is

$$\frac{hD}{k} = 0.385 \left(\frac{Cp}{k}\right)^{0.3} \left(\frac{DV\rho}{\mu}\right)^{0.56} \tag{10}$$

Each of the physical properties is evaluated at the film temperature t_f. By taking the value of Cp/k for air equal to 0.73, Equation 10 can be rewritten as

$$\frac{hD}{k} = 0.35 \left(\frac{DV\rho}{\mu}\right)^{0.56} \tag{11}$$

This equation is not recommended for values of $DV\rho/\mu$ less than 100; but for the range of values of velocity, temperature, and diameter of 0.1 to 50 lb/sec/sq ft, average 500°F, and 0.25 to 4.00 tube o.d., respectively, $DV\rho/\mu$ always exceeds this minimum.

For practical heat transfer work, one of the most reliable methods is that set forth by L. M. K. Boelter, V. H. Cherry, and H. A. Johnson in the University of California Syllabus Series on Heat Transfer. This method is expressed by the following empirical formula for air and gas:

$$h_0 = \frac{0.8(C)(Ta)^{1/3}(G_1)^{(0.60+0.08 \log d)}}{d^{0.53}} \tag{12}$$

where h_0 = outside film coefficient, Btu/(h)(sq ft)(F°)
C = constant (usually 1.25)
T_a = average gas temperature, degree absolute
G_1 = gas mass flow, pps/sq ft
d = tube o.d., in.

Equation 12 is the one we will use; it is presented graphically in Fig. 18, with temperature correction factors in Fig. 19.

The amount of heat transfer surface required in any equipment in which heat is to be transferred by convection from one fluid to another can be determined simply from Equation 13:

$$q = UA(\Delta t) \tag{13}$$

It is first necessary to determine (1) the rate at which heat is to be transferred, (2) the overall coefficient of convection U, and (3) the mean temperature difference Δt between the two fluids. Usually, in heat transfer equipment,

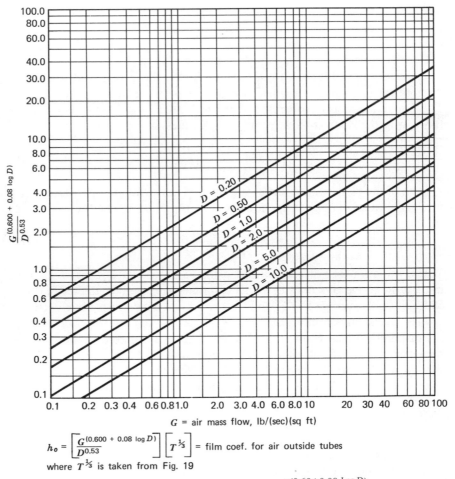

G = air mass flow, lb/(sec)(sq ft)

$$h_o = \left[\frac{G^{(0.600\ +\ 0.08\ \log D)}}{D^{0.53}} \right] \left[T^{\frac{1}{3}} \right] = \text{film coef. for air outside tubes}$$

where $T^{\frac{1}{3}}$ is taken from Fig. 19

Fig. 18. Mass flow–diameter factor versus mass flow. $\dfrac{G^{(0.60\ +\ 0.08\ \text{Log}\ D)}}{D^{0.53}}$ versus G for air for various tube diameters.

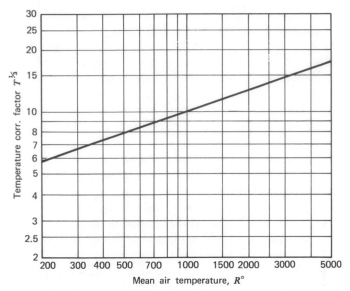

Fig. 19. Temperature correction factor for use with Fig. 18.

(Δt) refers to the log-mean temperature difference. The preceding three quantities may be determined as follows.

1. If the equipment is to be insulated, the rate of heat transfer q can be calculated from the properties of either fluid, since the rate at which the warmer fluid gives up its heat must be equal to the rate at which the cooler fluid absorbs heat. If the fluid used does not change phase, the rate of heat transfer q can be calculated from the rate of flow, the specific heat, and the temperature rise or fall of the fluid. If the fluid evaporates or condenses, the rate of heat transfer q can be calculated from the rate of flow and the heat of vaporization of the fluid. Any heat given up or absorbed by the fluid as a result of its being superheated or subcooled must also be included. In this chapter we consider only gas-to-gas conditions. Therefore, the equation for q is simply

$$q = (W^1)(\Delta t)(Cp) \qquad (14)$$

where q = Btu/h
$\quad W^1$ = fluid or gas flow, lb/h
$\quad \Delta t$ = temperature rise or fall, F°
$\quad Cp$ = average specific heat

Note that the temperatures and the rates of flow of the two fluids may not be chosen arbitrarily, since the values must be such that the heat balance is satisfied, and that at every section of the equipment the temperature of the warmer fluid must be higher than that of the cooler fluid. If neither fluid changes phase, and if the entering and leaving temperatures of the two fluids are specified, the rate of flow of either fluid may be assigned any value, but the rate of flow of the other fluid must be determined from the heat balance.

2.   The overall coefficient U can be calculated from the film coefficients of the two fluids as determined from Figs. 14–19 as applicable, including the coefficients for tube wall thickness and material, fouling factors, and a scaling factor (if applicable). The determination of the value of U for a particular case will be demonstrated later in this chapter.

The film coefficients for all types of convection depend on the temperature of the fluid. If the fluid does not change phase, its temperature changes as it flows through the equipment; consequently, the film coefficient also changes. A sufficiently accurate average value can usually be obtained, however, if the film coefficient is evaluated at the average temperature of the fluid, the average temperature being the arithmetic average of the temperature at which the fluid enters and leaves the equipment.

If the heat transfer surface consists of tubes, it may be necessary to select tentative values for tube sizes and numbers before the film coefficients can be determined. For example, the film coefficient for the fluid flowing through the tubes usually depends on the velocity of the fluid, which cannot be calculated until tube sizes and numbers have been selected. If these values are selected in advance, only the length of the tubes remains to be calculated after the amount of heat transfer surface required has been determined by the methods outlined previously. If the tube is either too long or too short, a new value for either tube size or number must be selected and the calculation repeated.

3.   The mean temperature difference between the two fluids must be used for Δt and can be calculated as explained previously.

Data on air flow indicate that approximate values of the film coefficient h for fluids flowing to banks of tubes can be determined by multiplying the value of h for flow normal to single tubes by 1.3, if the tubes are staggered, and by 1.2, if the tubes are in-line. Moreover, the average film coefficient of a bank of tubes increases as the number of rows increases.

Annular Spaces

The film coefficients for fluids flowing through the annular space of a double-pipe heat exchanger can be determined in the usual manner for fluids or gases flowing inside tubes, except that the diameter used to determine the diameter correction factor F_d is to be the equivalent diameter d_e.

$$d_e = \frac{d_1{}^2 - d_2{}^2}{d_2} \tag{15}$$

where d_e = equivalent diameter
$\quad\quad d_1$ = i.d. of the outer tube
$\quad\quad d_2$ = o.d. of the inner tube

Effect of Tube Diameter

Tube diameter affects both the outside and inside film coefficient. Research shows that the heat transfer coefficient varies inversely as D^m, where m ranges from 0.20 to 0.60. Work done at the University of California and proved in actual practice gives a value of $m = 0.53$.

Effect of Mass Flow

Gas mass flow has an exponential effect on the value of h, and h increases with $G_1{}^n$. The value of n is $(0.60 + 0.08^{\log d})$, as given by L. M. K. Boelter et al. This value again varies with different experimenters; however, the aforementioned value compares favorably with experimental data and therefore is recommended. A good approximate value of n is 0.606. This value is valid only for gas flow over the outside of tube bundles.

Effect of Gas Temperature

The effect of gas temperature upon h is still subject to disagreement. There is general agreement, however, that an increase in temperature increases h; most of the change can be attributed to gas radiation. The value of e in T^e can be taken as 0.33, with the value of gas temperature T in absolute Rankine units.

Packed Tubes

Apparent coefficients of heat transfer (based on the inside surface of the tube) for air flowing through vertical tubes packed with granular solids involve apparent mass velocities ranging from 900 to 14,000 lb of air/h (sq ft of cross section of tube). The ratio of the observed apparent coefficient h_A for the packed tube of diameter d_i to h_i for a 1-in. inside diameter empty tube, both the empty tube and packed tube having the same mass velocity based on gross cross section, depends on the ratio d_p/d_i, where d_p is the diameter of the packing. Table 5 gives various values of d_p/d_i and h_A/h_i.

TABLE 5   Values of d_p/d_i and h_A/h_i

d_p/d_i	0.05	0.10	0.15	0.20	0.25	0.30
h_A/h_i	5.5	7.0	7.8	7.5	7.0	6.6

Gases Heated or Cooled Inside Coils

For gases in turbulent flow inside coiled tubes or pipes, the following equation can be used:

$$h_{\text{coils}} = \left[1 + 3.54 \left(\frac{D}{D_c} \right) \right] (h_{\text{tubes}}) \tag{16}$$

where D = inside diameter of pipe or tube, in.

$\quad\quad D_c$ = diameter of coil or helix, in.

$\quad\quad h_{\text{tubes}}$ = coefficient for gases inside straight tubes determined in usual way

Tests on air cooled in two helical coils of $1\frac{1}{4}$-in. seamless steel tubing, one having a diameter of 24.8 in. and two turns, and the other having a diameter of 8.27 in. and six turns, with flow always turbulent and DG up to 150,000, led to the proposal of the following equation:

$$\frac{hD}{k} = \left[0.039 + \frac{0.138D}{D_c} \right] \left[\frac{DG\,C_p}{k} \right]^{0.76} \tag{17}$$

where D/D_c = ratio of the tube diameter to the coil or helix diameter. For ordinary use, however, simply multiply the value of h for straight pipe by $[1 + 3.5\,(D)/(D_c)]$. Therefore, for a given Reynolds number, h is higher for coiled pipes than for straight pipes. The pressure drop in straight pipes is also higher, however, because of the increased turbulence in coiled pipes.

Fouling and Roughness

Studies exist of the cooling of air in tubes with three degrees of artificial roughness, in which the height of the roughness pyramid ranges from 1/45 to 1/7 of the radius of the pipe. Although in the turbulent region friction was as high as six times that for smooth tubes, heat transfer was only 10 to 20% greater than for smooth tubes. It was concluded that for the same power loss or pressure drop, greater heat transfer was obtained from a smooth than from a rough tube.

The effect of a given scale deposit is usually far less serious for a gas than for water because of the higher thermal resistance of the gas film compared with that of the water film. However, layers of dust or of materials that sublime, such as sulfur, may seriously reduce heat transfer between gas and solid. Thus, because of thick deposits of dust in the flues of blast furnace stoves, the overall coefficient of heat transfer reduced approximately 40%; for clean surfaces, the values of U were roughly double those computed in clean pipes.

The determination or estimation of the fouling factor is always a problem because there are so many variables in this determination. Dust content of the gas, deposition of condensables, gas velocity, tube roughness, and electro-static charge of the dust solids in relation to the tube wall all affect fouling factor. The following fouling factors listed in Table 6 can generally be assumed.

TABLE 6   Proposed Fouling Factors

Fouling Factors	R_f
Relatively clean gas	0.0005
Air containing oil vapor	0.0025
Sewage sludge incinerator gas	0.0030
Refuse incinerator flue gas	0.0035
Gas-fired gas turbine exhaust	0.0010
Oil-fired gas turbine exhaust	0.0015
Waste liquid incinerator exhaust gas	0.0015
Fume incinerator exhaust gas	0.0010
Reciprocating engine exhaust, oil fired	0.0030
Reciprocating engine exhaust, gas fired	0.0020

These values of R_f are to be used as one factor in the final determination of U.

Since the formation of scale in a gas-to-gas heat exchanger is not common, we deal with determination of the scale factor in a later chapter.

Natural Convention to Tubes

Natural convection results from mixing one fluid with another, the motion of the fluid being caused by differences in density resulting from temperature difference. Energy is also transferred simultaneously by molecular conduc-tion and, in transparent media, by radiation.

Most heat transfers from gas-to-gas heat exchange are in a forced convec-tion or turbulent flow. Therefore, we deal only briefly with the subject of natural convection, setting forth a workable method of problem solution. The natural convection heat transfer coefficient can be determined by the following equation:

$$h = (h_0)(F_t)(F_d)(F_p) \tag{18}$$

where h = film coefficient, Btu/(sq ft)(h)(F°)
     h_0 = base value of coefficient from Fig. 20
     F_t = temperature correction factor from Fig. 21
     F_d = diameter correction factor from Fig. 21
     F_p = pressure correction factor from Fig. 21

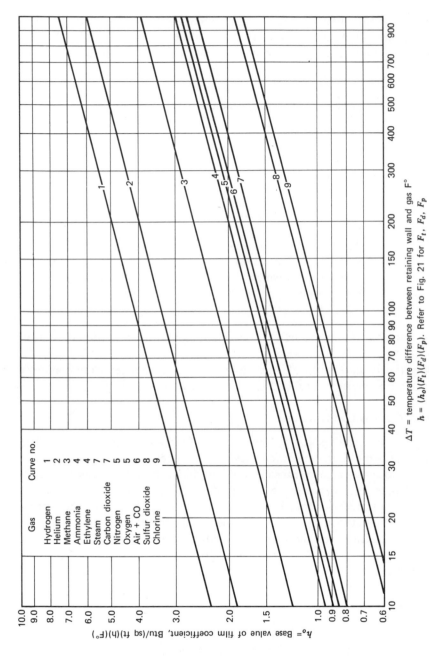

Fig. 20. Base value of the film coefficient for gases heated outside horizontal tubes, natural convection. From Applied Heat Transmission, Stoever.

where Fig. 20 illustrates the base value of the heat transfer coefficient for gas heated outside horizontal tubes, natural convection, and Fig. 21 shows the convection factors for temperature, diameter, and pressure.

Note that the film coefficient h obtained from Equation 18 is for heat transfer by convection only. If the heat transfer surface is exposed to other surfaces at different temperatures, it will gain or lose heat by radiation also.

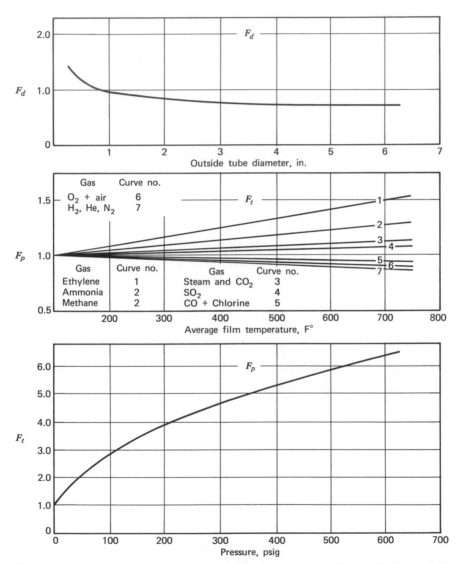

Fig. 21. Tube diameter, temperature, and pressure correction factors F_d, F_t, and F_p for Fig. 20. From Applied Heat Transmission, Stoever.

The film coefficient for the combined effect of convection and radiation can be obtained by adding the product of the emissivity of the surface times the radiation coefficient h_r and the convection coefficient h obtained from Equation 18.

The following is a list of applications involving natural convection between tubes or pipes and a fluid.

1. Fire tube boilers, where the boiling water surrounding the fire tubes is in natural convection.
2. Evaporators, where the condensing vapor is in natural convection in relation to the condensing surface.
3. Jacketed kettles (without mechanical agitation).
4. Heating coils immersed in tanks of liquid. The heating coil or element may be a steam coil, an electrical immersion heater, or a fired immersion pipe.

Gases Heated on Vertical Plates, Natural Convection

The natural convection heat transfer coefficient to vertical plates may be determined by the equation

$$h = (h_0)(F_t)(F_p) \tag{19}$$

where h = film coefficient, Btu/(sq ft)(h)(F°)
$\quad\ F_t$ = temperature correction factor from Fig. 23
$\quad\ F_p$ = pressure correction factor from Fig. 23

Figure 22 illustrates the base value of the heat transfer coefficient for gases heated outside vertical plates or vertical tubes, natural convection, and Fig. 23 shows correction factors for temperature and pressure.

The film coefficient h obtained from Equation 19 is for convection only. If the heat transfer surface is also subject to radiation, this must be added to h by the factor h_r multiplied by the emissivity of the surface.

A simplified determination of natural convection is expressed by the following:

Horizontal plates facing upward, $h = 0.38 \, (\Delta t)^{0.25}$
Horizontal plates facing downward, $h = 0.20 \, (\Delta t)^{0.25}$
Vertical plates more than 1 ft high, $h = 0.27 \, (\Delta t)^{0.25}$
Vertical plates less than 1 ft high, $h = 0.28 \left(\dfrac{\Delta t}{L}\right)^{0.25}$

where

L is in ft

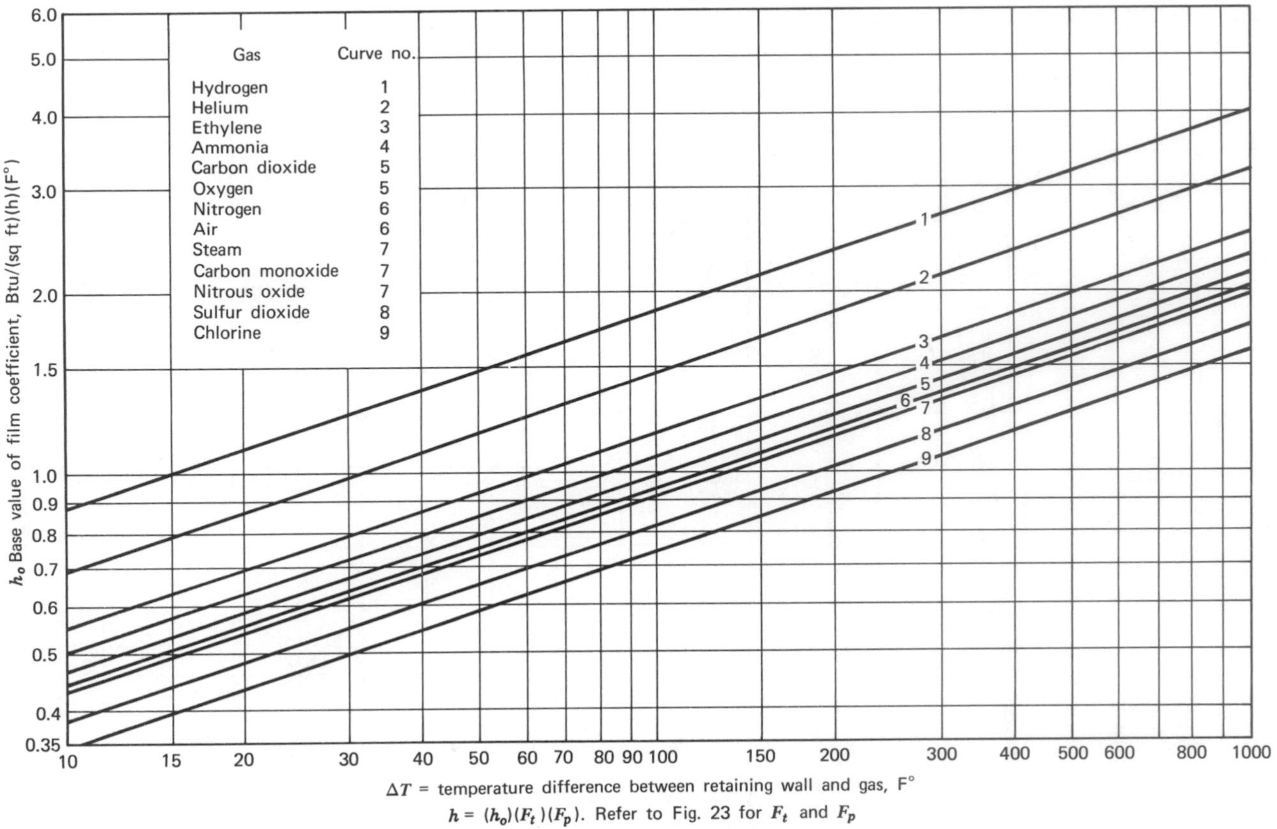
Gas
Curve no.
Hydrogen 1
Helium 2
Ethylene 3
Ammonia 4
Carbon dioxide 5
Oxygen 5
Nitrogen 6
Air 6
Steam 7
Carbon monoxide 7
Nitrous oxide 7
Sulfur dioxide 8
Chlorine 9
h_o Base value of film coefficient, Btu/(sq ft)(h)(F°)
6.0
5.0
4.0
3.0
2.0
1.5
1.0
0.9
0.8
0.7
0.6
0.5
0.4
0.35
10
15
20
30
40
50
60
70
80
90
100
150
200
300
400
500
600
800
1000
1
2
3
4
5
6
7
8
9
ΔT = temperature difference between retaining wall and gas, F°
$h = (h_o)(F_t)(F_p)$. Refer to Fig. 23 for F_t and F_p

Fig. 22. Gases heated inside or outside vertical tubes or on vertical plates, natural convection. From Applied Heat Transmission, Stoever.

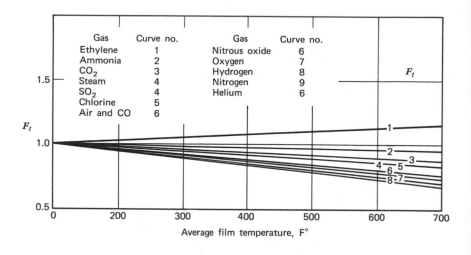

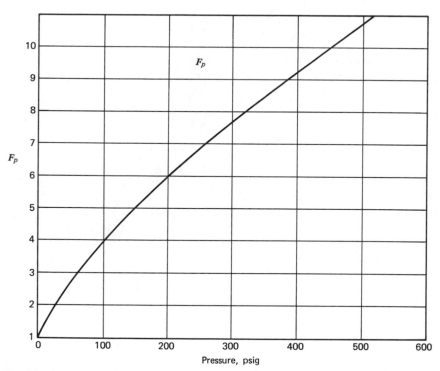

Fig. 23. Correction factors for temperature and pressure F_t and F_p for use with Fig. 22. From Applied Heat Transmission, Stoever.

Another expression for the film coefficient for air heated on horizontal plates, natural convection, for flat plates of 3 sq ft or more is

$$h_H = 1.27 \, (h_v) \text{ plate facing upward}$$
$$h_H = 0.67 \, (h_v) \text{ plate facing downward}$$

where h_H = coefficient for horizontal plates
h_v = coefficient for vertical plates based on Equation 19 and associated curves and data

Note that the film coefficients obtained by the preceding method are for convection only. If the plate surface is exposed to other surfaces at different temperatures, it will gain or lose heat by radiation also. The film coefficient for the combined effect of convection and radiation can be obtained by adding the product of the emissivity of the surface times the radiation coefficient h_r and the convection coefficient h obtained by the preceding method.

Examples of natural convection to plates are

1. Plate coils for tank heating.
2. Heated tank walls, where the surrounding cool air is in natural convection with respect to the hotter tank wall. (Some radiation is also involved here.)

Determination of Tube Temperature

The determination of tube wall temperature is an important design factor because it largely determines the selection of tube material.

The heat transfer methods detailed in the foregoing pages all involve either the temperature difference Δt between the surface of the retaining wall and the main body of the fluid or the film temperature t_f, which is arbitrarily defined as the arithmetic average of the temperature of the surface of the retaining wall and the temperature of the main body of the fluid. In order to evaluate either of these quantities, the temperature of the surface of the retaining wall must be known. The assumption of a value of tube wall temperature is sometimes sufficient for estimations. For more accurate results, the exact temperature of the retaining wall can be calculated by trial and error, as follows.

Refer to Fig. 24, Temperature Gradient Through Retaining Wall. Neglecting the temperature drop through the retaining wall itself and using the nomenclature shown on Fig. 24, (1) a tentative value is assumed for the temperature t_w of the retaining wall (tube wall). (2) Based on this value, the values of Δt_1 and Δt_2 and t_f are calculated. (3) The film coefficients h_1 and h_2 corresponding to these values are determined from the methods discussed

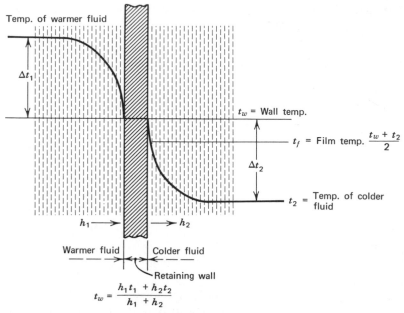

Fig. 24. Diagram of temperature gradient through two fluids and retaining wall.

earlier. Using these values of h_1 and h_2, a more accurate value of t_w is calculated by the equation

$$t_w = \frac{h_1 t_1 + h_2 t_2}{h_1 + h_2} \qquad (20)$$

This value of t_w is usually sufficiently accurate and represents the average
value of the tube or retaining wall temperature. Using Equation 20, values
of Δt_1, Δt_2, and t_f can again be calculated and the final value of h_1 and h_2
can then be determined.

Equation 20 is based on the simplifying assumption that the temperature
drop through the retaining wall is negligible and that the areas of the two
surfaces of the wall are approximately equal. Therefore,

$$q = h_1 A(t_1 - t_w) = h_2 A(t_w - t_2) \qquad (21)$$

It is often necessary and desirable to know the value of t_w across a retaining
wall or tube wall to calculate thermal stress and film temperatures of critical
fluids such as organic heat transfer fluids. To determine Δt_w, q/A must be
known; it can be determined by the foregoing methods. The following equation may also be used:

$$q/A = \Delta t_w k_w \frac{(2\pi l)}{\left[\log_\varepsilon \frac{d_o}{d_i} \right]} \qquad (22)$$

where q/A = maximum heat transfer rate, Btu/sq ft
    Δt_w = tube wall temperature gradient
    k_w = thermal conductivity of tube wall material, Btu/(h)(sq ft)(F°)(ft)
    l = length of tube, ft per sq ft of heating surface
    d_o = tube o.d., in.
    d_i = tube l.d., in.

Equation 22 may be rewritten

$$\Delta t_w = k_w \frac{\dfrac{q/A}{(2\pi l)}}{\left[\log_\varepsilon \dfrac{d_o}{d_i}\right]} \tag{23}$$

GAS-SIDE PRESSURE DROP

The determination of gas-side pressure drop in designing gas-to-gas heat recovery equipment is important because this must be used in selection of blowers, sizing of ducts, and the like. In all the following discussion of gas pressure drop, we assume constant density, which is not actually true because gas density varies inversely with absolute temperature. However, the density chosen will be at the average temperature of the gas in question.

Gas Pressure Drop Inside Smooth Tubes, Turbulent Flow

For isothermal turbulent flow of fluids inside straight smooth tubes, pressure drop per foot can be calculated by the equation

$$\frac{\Delta P}{N} = \frac{4f \rho V^2}{2gD} \tag{24}$$

The friction factor f experimentally for the range of $DV\rho/\mu$ equal to 4000 to 1,000,000 may be expressed by the following equation:

$$f = \frac{0.0653}{\left(\dfrac{DV\rho}{\mu}\right)^{0.228}} \tag{25}$$

This equation is valid only for isothermal flow. However, it may be used as an approximation for the heating and cooling of any gas because the viscosities of gases do not change rapidly with temperature change.

The curves representing values of $(\Delta P/N)_0$ for several gases is based on Equation 25 and is illustrated by Fig. 25. Diameter and temperature correction factors are illustrated by Fig. 26.

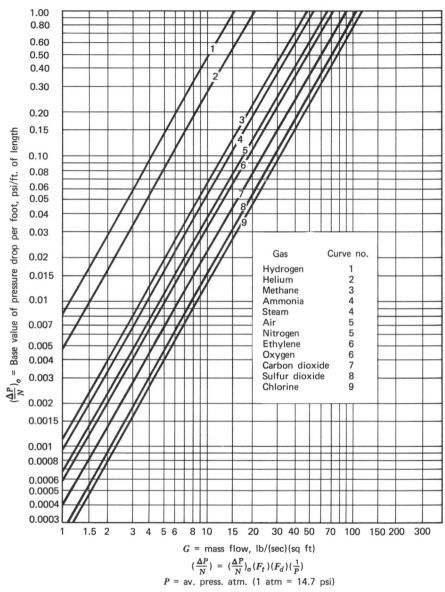

Fig. 25. Base value of pressure drop, gases inside smooth tubes, turbulent flow. From Applied Heat Transmission, Stoever.

The chart shows the following:

Y-axis: $\left(\frac{\Delta P}{N}\right)_o$ = Base value of pressure drop per foot, psi/ft. of length

X-axis: G = mass flow, lb/(sec)(sq ft)

$$\left(\frac{\Delta P}{N}\right) = \left(\frac{\Delta P}{N}\right)_o (F_t)(F_d)\left(\frac{1}{P}\right)$$

P = av. press. atm. (1 atm = 14.7 psi)

Gas	Curve no.
Hydrogen	1
Helium	2
Methane	3
Ammonia	4
Steam	4
Air	5
Nitrogen	5
Ethylene	6
Oxygen	6
Carbon dioxide	7
Sulfur dioxide	8
Chlorine	9

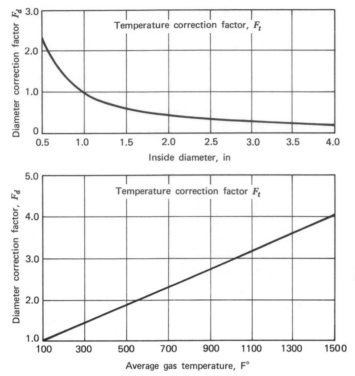

Fig. 26. Correction factors for diameter and temperature F_d and F_t for use with Fig. 25. From Applied Heat Transmission, Stoever.

Based on the data of Figs. 25 and 26, the equation for gas pressure drop inside tubes is

$$\left(\frac{\Delta P}{N}\right) = \left(\frac{\Delta P}{N}\right)_0 (F_t)(F_d)\left(\frac{1}{P}\right) \tag{26}$$

where $\dfrac{\Delta P}{N}$ = pressure drop, psi/ft

$\left(\dfrac{\Delta P}{N}\right)_0$ = base value of pressure drop, psi from Fig. 21

F_t = temperature correction factor from Fig. 21

F_d = tube diameter correction factor from Fig. 21

p = average gas pressure, atm absolute (1 atm = 14.7 psi)

Pressure Drop Inside Coiled Pipes

For isothermal flow in curved or coiled pipes, the pressure drop may be considerably more than in straight pipes, conditions being otherwise the same. For isothermal streamline flow in curved pipes, available data show the flow mechanism and allow calculations of pressure drop. Color band work shows that the fluid particles follow tortuous paths, traveling from the center of the pipe toward the outside wall and then crossing back toward the inside wall. Near the wall, the fluid particles travel faster in a curved section of pipe than in a straight section because of their spiral path. Because of this increased velocity near the wall and the longer path traveled per foot of pipe by individual fluid particles, a higher pressure drop in curved pipes is to be expected. Tests also show that streamline flow can exist at much higher Reynolds numbers in curved pipe than in straight pipe.

Gas pressure drop for coiled pipes can be determined in the same manner as for straight pipes or tubes using Figs. 25 and 26. The total actual pipe length is then increased by the amount determined from Fig. 27 for 90° bends with adjustment for complete 360° bends, as stated on the figure.

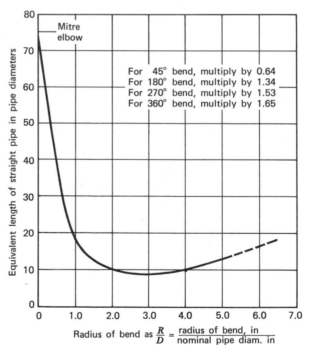

Fig. 27. Curve for converting 90° bends into equivalent straight pipe.

Pressure Drop in Annular Spaces

For turbulent flow, pressure drop in rectangular ducts and annular spaces may be determined in the same way as for a circular pipe, using an "equivalent diameter" equal to $4r_h$, where r_h is the hydraulic radius $r_h = S/b$, where S is the cross section normal to flow, in sq ft, and b is the breadth of the wetted perimeter, in ft. This equivalent diameter is then used to determine the diameter correction factor.

Gas Pressure Drop at the Entrance to Tubes

Figure 28 is based on the values of the contraction coefficient K given in Table 7.

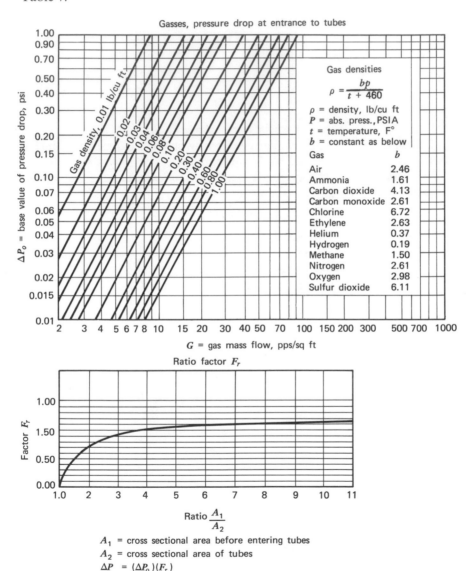

Gasses, pressure drop at entrance to tubes

Gas densities

$$\rho = \frac{bp}{t + 460}$$

ρ = density, lb/cu ft
P = abs. press., PSIA
t = temperature, F°
b = constant as below

Gas	b
Air	2.46
Ammonia	1.61
Carbon dioxide	4.13
Carbon monoxide	2.61
Chlorine	6.72
Ethylene	2.63
Helium	0.37
Hydrogen	0.19
Methane	1.50
Nitrogen	2.61
Oxygen	2.98
Sulfur dioxide	6.11

G = gas mass flow, pps/sq ft

Ratio factor F_r

A_1 = cross sectional area before entering tubes
A_2 = cross sectional area of tubes
$\Delta P = (\Delta P_o)(F_r)$

Fig. 28. ΔP entrance to tubes.

TABLE 7   Contraction Coefficient for
              Sharp Edged Entrances

Ratio of Larger to Smaller Area	K
1.0	0.00
1.5	0.13
2.0	0.19
2.5	0.24
3.0	0.27
3.5	0.30
4.0	0.32
6.0	0.38
8.0	0.41
10.0	0.43

The use of Fig. 28 assumes that gas density changes only slightly during contraction, which is permissible for moderate rates of flow.

The pressure drop at the entrance to unflared tubes (or at any other abrupt contraction in cross section) can be determined by

$$\Delta p = \Delta p_0 \ (F_r) \tag{27}$$

and values of F_r from Table B.

where Δp = pressure drop, psi
Δp_0 = base value of pressure drop, psi, from Fig. 28
F_r = area-ratio correction factor from Table 8

TABLE 8   Area Ratio versus Correction
              Factor [a]

$\dfrac{A_1}{A_2}$	F_r
1.0	0.00
1.5	0.54
2.0	0.75
2.5	0.86
3.0	0.92
3.5	0.97
4.0	1.00
6.0	1.07
8.0	1.11
10.0	1.13

[a] A_1 = cross-sectional area before entering the tubes
A_2 = cross-sectional area of the tubes

TABLE 9   Gas Densities

Gas	b
Acetylene	2.46
Air	2.70
Ammonia	1.61
Butane	5.57
Carbon dioxide	4.13
Carbon monoxide	2.61
Chlorine	6.72
Ethane	2.83
Ethylene	2.63
Helium	0.37
Hydrogen	0.19
Hydrogen sulfide	3.21
Methane	1.50
Methyl chloride	4.82
Nitrogen	2.61
Oxygen	2.98
Sulfur dioxide	6.11
Nitrous oxide	4.13

Approximate gas densities may be calculated by the equation

$$p = \frac{bp}{t + 460} \text{ and values of } (b) \text{ from table 9} \tag{28}$$

where p = density of gas, lb per cu ft
$\qquad p$ = absolute pressure of gas, lb per sq in. absolute
$\qquad t$ = gas temperature, F°
$\qquad b$ = a constant obtained from Table 9

Gas Pressure Drop in Tube Banks

The determination of gas pressure drop in tube banks involves more variables than gas pressure drop inside tubes. The following equation can be used:

$$\Delta P = \left[\frac{(f)(G^2)(N)(\lambda)}{6 \times 10^{10}(D_{ev})(\rho)} \right] \left[\frac{\left(\frac{12 D_{ev}}{Sy} \right)^{0.4} \left(\frac{Sx}{Sy} \right)^{0.6}}{\left(\frac{\mu}{\mu_w} \right)^{0.14}} \right] \tag{29}$$

where ΔP = pressure drop, psi
$\qquad f$ = friction factor from Fig. 29
$\qquad G$ = mass flow, lb/h(sq ft)
$\qquad N$ = number of rows (passes)

λ = flow path per row, ft
D_{ev} = volumetric equivalent diameter, ft
ρ = gas density (average) lb per cu ft
μ = bulk viscosity, lb/(ft)(h)
μ_w = viscosity at wall temperature, lb/(ft)(h)
Sy = center-center distance between adjacent tubes in same row, in.
Sx = center-center distance between adjacent tubes in adjoining rows, in.

$$D_{ev} = \frac{4NFV}{A_t}, \text{ ft}$$

NFV = net free volume, cu ft per ft
A_t = total sq ft of surface per lineal ft of tube

$$NFV = \frac{SxSy - \pi d_o\,(do/4)}{144} \text{ cu ft}$$

d_o = tube outside diameter.

$$G \quad = \frac{W}{NFA}, \text{ lb/(h)(sq ft)} \tag{31}$$

where NFA = net free area of tube bank, sq ft

See Fig. 30 for a diagram of tube arrangement and illustration of pertinent dimensions.

Gas pressure drop in equipment of this type is usually expressed in inches of water column. To obtain pressure drop in inches of water, multiply the preceding result in psi by 27.7. Equations 29, 30, and 31 are also used for calculating gas pressure drop through banks of finned tubes by substituting A_t as the total external surface of the finned tube.

AN EXAMPLE OF GAS-TO-AIR HEAT RECOVERY

A practical example of gas-to-air heat recovery is the application of a tubular recuperator in preheating combustion air by hot combustion products from a boiler, an incinerator, or any other source of combustion gas. Figure 31 illustrates overall arrangement and dimensions.

In this example, assume that the combustion products are from a sewage sludge incinerator. Typical products from such an incinerator are as follows:

Total combustion products—40,000 lb/h
Dry gas total—29,000 lb/h
Water vapor total—11,000 lb/h
Percentage of water vapor—27.5
Temperature—1400°F

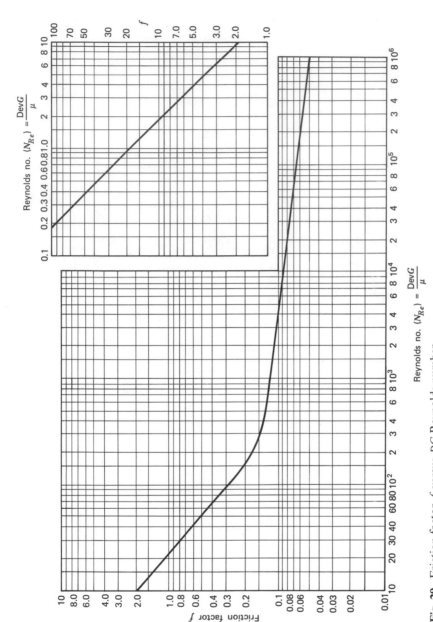

Fig. 29. Friction factor f versus DG Reynolds number.

61

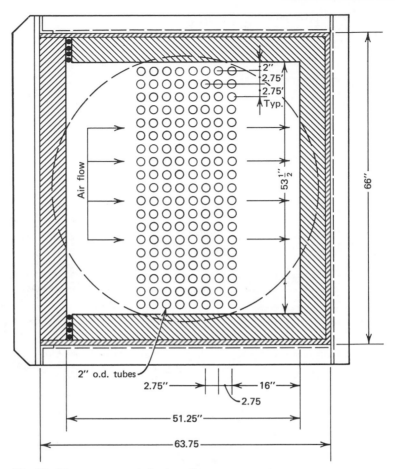

Fig. 30. Heat recovery air heater tube arrangement.

Using the heat source above, heat combustion air as follows:

Combustion air rate—19,000 lb/h
Air temperature into unit—70°F
Air temperature out of unit—1000°F
Allowable tube side pressure drop—9 in./wc (water column)
Allowable shell side pressure drop—4 in./wc
Heat recovery rate = (19,000)(0.25)(1000−70)
$$q = 4,417,500 \text{ Btu/h}$$

This heat rate may be verified by using the tables of Thermodynamic Properties of Air by Keenan and Kay.

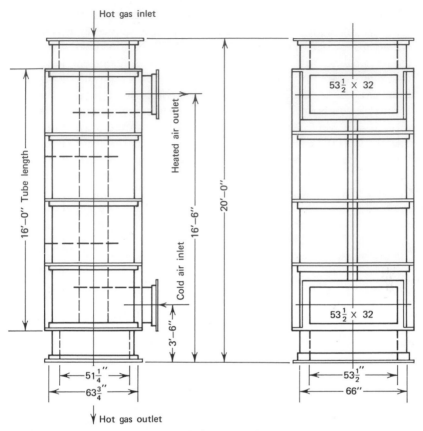

Fig. 31. Outline drawing, sample tubular air-gas heat recovery unit.

Heat content of air, 1000°F = 263.34 Btu/lb
Heat content of air, 70°F = 31.17 Btu/lb
Heat gain   = 232.17 Btu/lb

Agreement is sufficiently close for practical purposes.

The specific heat of combustion gas is taken from Fig. 6. However, a reasonably accurate final temperature must be known. As a first step, an estimate of specific heat must be made. Here, take a value of 0.35.

$$\text{Combustion gas } \Delta t = \frac{4,417,500}{(40,000)(0.35)} = 316°F$$

Estimate of final gas temperature = 1400 − 316 = 1084°F

Use this final gas temperature estimate (1084°F) plus the initial gas temperature (1400°F) = 2484°F on Fig. 6 on the curve for water vapor content of 27.5% and find specific heat of 0.35. Since the checked value of specific heat of 0.35. Since the checked value of specific heat coincides with the original value, no further trials are needed. If it had not coincided, several trials would have had to be made until agreement or near agreement was reached.

A straight-tube unit is used with one gas pass on the tube side and with multiple gas passes on the shell side. The hot combustion gas flows inside the tubes and the combustion air on the outside (shell side). This flow path is used because (1) straight tubes are easy to clean by running a mechanical tube cleaner through the tubes, and (2) with the cooler gas outside the tubes, the shell construction can be less expensive. Tube temperature is 1400°F at one end and 1084°F at the other; therefore, from Table 3, the tube material should be stainless steel of the 300 series, such as type 304 or 312.

The coefficients of convection outside and inside will be comparable; therefore it is not advantageous to use extended surface (fins) on either the outside or the inside of the tubes.

The log-mean temperature difference for this case is

$$\text{Lm } \Delta T = \frac{(1084 - 70) - (1400 - 1000)}{2.3 \log_{10}\left(\dfrac{1084 - 70}{1400 - 1000}\right)} = 661°F$$

From Fig. A-h in the Appendix, the correction factor for Lm ΔT is 0.98.

$$\text{Final Lm } \Delta T = (661)(0.98) = 648°F$$

At this point, the tube size must be selected. If the gas flowing inside the tubes were clean, then small-diameter tubes could be used (1 in. or less), and less heating surface would be required. Since we are handling dirty combustion gas from a sewage sludge incinerator containing a substantial quantity of dust and ash, a tube size should be selected that can be easily cleaned by air-driven cleaning brushes. Generally, when a sewage sludge incinerator is operated at exhaust gas temperatures above 1200°F, all hydrocarbons are consumed and no sticky material will be in the gas stream; except for inadvertent periods of hydrocarbon carryover, only dust and ash should be present.

Tube wall thickness is not determined in this case by operating pressure but by the need to provide adequate structural integrity and wear allowance. Therefore, for this example, select 2.00-in. o.d., 14 gauge (0.083) wall-type 304 stainless steel, either welded or seamless.

Tube i.d. = 1.834 in.
Inside heating surface, per ft   = 0.480 sq ft
Inside transverse area per tube = 2.642 sq in.
                                              = 0.0183 sq ft

Next, determine inside gas mass flow. Mass flow is limited by the allowable gas pressure drop on the tube side, which cannot be finally determined until the effective heat exchanger length is known. Therefore, a mass flow value must be taken for a trial calculation. If the first choice results in too high or too low a gas pressure drop, then lesser or greater trial must be made, depending on whether the calculated gas pressure drop is greater or less than the stated allowable value.

Try internal mass flow, $G_1 = 4.0$ pps/sq ft

Gas flow in tubes $= 40,000$ lb/hr $= 11.11$ pps

Required cross section $= \dfrac{11.11}{4.0} = 2.780$ sq ft

Number of tubes $= \dfrac{2.780}{0.0183} = 152$ tubes

Inside surface per ft of tube bundle $= (152)(0.480) = 72.96$ sq ft

From Fig. 14, for $G = 4.0$, $h_{io} = 12.0$ (for air)

From Fig. 15, $F_t = 1.40$ for average temperature, $1242°F$

　　　　　　$F_d = 0.90$ for 1.834 i.d.

From Equation 8,

$$h_i = (h_{io})(F_t)(F_d) = 15.12 \text{ Btu/(h)(sq ft)(F°)}$$

The shell side air flow will be in as many passes as possible consistent with the air pressure drop allowance. With 152 parallel tubes arranged in a rectangular pattern, the following combinations are possible; (1) 8×19, (2) 10×15, or (3) 7×22. Here again, an arbitrary selection must be made, and the 8×19 arrangement is chosen, with 19 tubes normal to the direction of gas flow, and the tube bundle 8 tubes deep.

Arrange the tubes on 2.75-in. centers, with $\frac{5}{8}$ in. average clearance between the outside tubes in the row and the shell inside wall.

The through flow (gas passage) area per ft of tube bundle is

$$\frac{(19 - 1)(0.75)(12) - (2)(0.625)(12)}{144} = 1.02 \text{ sq ft}$$

There will be many tube rows in the air path and several reversals of flow. Therefore, the mass flow must be comparatively low so as not to exceed the allowable gas pressure drop.

Take the external (shell side) mass flow rate, at $G_2 = 2.0$ pps/sq ft

Combustion air flow $= 19,000$ lb per h $= 5.28$ pps

Required gas passage area per pass $= \dfrac{5.28}{2.0} = 2.64$ sq ft

Average length per pass $= \dfrac{2.64}{1.02} = 2.59$ ft

From Equation 12,

$$h_o = \frac{(0.80)(C)(T_a^{1/3})(G_2)^{(0.60+0.08 \log d)}}{d^{53}}$$

where C = 1.25
 T_a = (1000 + 70)/(2) + 460 = 995°R
 G_2 = 2.00 pps, sq ft
 d = 2.00 in.
 h_o = 10.67 Btu/(sq ft)(F°)(h)

Tube wall $h_w = \dfrac{k}{L}$

where k = thermal conductivity of tube = 16
 L = wall thickness, ft = 0.00691
 h_w = 2315 Btu/(h)(sq ft)(F°)

$R_i \;= \dfrac{1}{h_i} = \dfrac{1}{15.12} = 0.0666138$

$R_o \;= \dfrac{1}{h_o} = \dfrac{1}{10.67} = 0.0937053$

$R_w \;= \dfrac{1}{h_w} = \dfrac{1}{2315} = 0.0004319$

R_{fi} = fouling resistance, inside from Table 6 = 0.003
R_{fo} = fouling resistance, outside from Table 6 = 0.0005
$R_t \;= R_i + R_o + R_w + R_{fi} + R_{fo} = 0.1643$

$U_o \;= \dfrac{1}{R_t} = 6.08$ Btu/(h)(sq ft)(F°)

Required heating surface $= S = \dfrac{q}{(U_o)(\text{Lm } \Delta T)}$

$\qquad\qquad = \dfrac{4,417,500}{(6.08)(648)} = 1121$ sq ft

From previous calculations, the effective heating surface per foot of tube bundle is 72.96 sq ft. The effective tube length = 1121/72.96 = 15.37 ft.
 Make tube length 16 ft. Total effective heating surface is 1167.36 sq ft.
 Number of shell side passes = 16/2.59 = 6.

The heat exchanger consists of 152 tubes 16 ft long, 2-in. o.d., 0.083 wall, type 304, stainless steel in a shell of steel plate with internal insulation, and baffled to provide six air passes. However, to know whether this configuration can be used, the pressure drops on both sides must first be determined.

To determine hot gas pressure drop through the tubes, take the following:

Flow rate = 40,000 lb/h

Water vapor = 11,000 lb/h

Gas temperature into tubes = 1400°F

Gas temperature out of tubes = 1084°F

Net flow area (152 tubes) = 2,782 sq ft

Mass flow $\dfrac{40,000}{(3600)(2.782)}$ = 4.0 pps (sq ft)

Tube I.d. = 1.834 in.

Average gas temperature = $\dfrac{1400 + 1084}{2}$ = 1242°F

From Fig. 25, for G = 4.0 $\dfrac{(\Delta P)}{(N_o)}$ = 0.0075 for air

From Fig. 26, values of F_d and F_t are as follows:

For d = 1.834, F_d = 0,460

For T = 1242°F, F_t = 3,420

$\dfrac{\Delta P}{N}$ = (0.0075)(0.460)(3.420) = 0.011799 psi per ft

ΔP = (0.011799)(16) = 0.187505 psi

Now consider pressure loss at the entrance to the tubes. From Fig. 28, for G = 4.0 and p = 0.023,

$$\Delta P_o = 0.098$$

$$\frac{A_1}{A_2} = \frac{\text{Cross section before tubes}}{\text{Tube cross section}} = \frac{2730}{2.782} = 981$$

This value of A_1/A_2 is out of the range of Table 8. Take F_r = 1.20

$$\Delta P_e = (\Delta P_o)(F_r) = (0.098)(1.20) = 0.1176 \text{ psi}$$

Total internal gas pressure drop = $P + P_e$ = 0.1975 + 0.1176 = 0.3051 psi = 8.45 in. w.c.

The tube side pressure drop is therefore within the limits stated. The actual limits of pressure drop must be determined for each case. In many instances, this limit may be between 4 and 6 in., with consequent differences in value of G.

The pressure drop on the shell side must be calculated by the use of Equations 28, 29, and 30. As stated before, the tubes will be arranged in a rectangular pattern, 8 tubes by 19 tubes spaced 2.75 in. center-center, with 19 tubes normal to the gas flow, and 6 shell side passes.

Combustion air flow—19,000 lb/h
Temperature in—70°F
Temperature out—1000°F

With a 16-ft tube length and 6 shell side passes, the average length per pass is 2.67 ft.

This arrangement produces a through flow net area of 2.73 sq ft.

$$G \quad = \frac{19,000}{2.73} = 6960 \text{ lb/(hr)(sq ft)}$$

$$N \quad = 8 \times 6 = 48 \text{ tube passes}$$

$$= \text{flow path per row} = \frac{2.75}{12} = 0.229 \text{ ft}$$

$S_x \quad = 2.75 \text{ in.}$
$S_y \quad = 2.75 \text{ in.}$
$A_t \quad = 0.524 \text{ sq ft } (l \text{ ft}) \text{ of tube}$
$d_o \quad = \text{tube o.d., 2 in.}$

$$NFV = \left[\frac{(2.75)(2.75) - \pi \left(2.0 \left(\frac{2.0}{4} \right) \right)}{144} \right] = 0.0307$$

$$D_{ev} \quad = \frac{4NFV}{A_t} = \frac{4(0.0307)}{0.527} = 0.234$$

$\rho \quad = 0.0387 \text{ lb/(cu ft) at average tempertaure 535°F}$
$\mu \quad = 0.0666 \text{ lb/(ft)(h) at 535°F (From Appendix, Table B-1 Values of Viscosity)}$
$\mu_w \quad = 0.0360 \text{ lb/(ft)(h) at 978°F wall temperature}$
$$\frac{D_{ev}G}{\mu} = \frac{(0.234)(6960)}{0.0666} = 24,454$$

From Fig. 29, friction factor $f = 0.077$.
Substituting these values in Equation 28,

$$\Delta P = 0.0755 \text{ psi}$$
$$= 2.09 \text{ in. w.c.}$$

Six gas passes require five 180° flow reversals. Consider each 180° reversal as a ducted return having a cross section of approximately 52 in. by 15 in. = 5.42 sq ft.

The average value of ρ = 0.0387 lb/cu ft
Air flow = 8.88 cu ft/min

$$\text{Average air velocity} = \frac{8188}{5.42} = 1510 \text{ fpm}$$

A 180° return with short radius equals approximately 36 diameters of straight duct. Five returns equal approximately 180 diameters of straight duct.

The equivalent diameter of 52 × 15 in. equals 32 in. Therefore, the equivalent length = (32/12)180 = 480 ft of straight duct.

From duct pressure loss chart, Appendix Chart No., K,

$$\Delta P = (0.073)(4.8)(.074)/(0.387) = 0.67 \text{ in. } (.16)\left(\frac{480}{100}\right)\left(\frac{.074}{.0387}\right) = 1.47$$

Total air-side pressure drop = 2.09 + 1.47 = 3.56.

Note that the allowable pressure drop on the shell side is stated as 4 in. water column. The result above is 3.56 in., which is below the allowable. If air pressure drop is greater, approaching the 4-in. limit, another trial must be made, selecting an air mass flow rate slightly higher than the 2 pps per sq ft of this example. Pressure drop varies directly as the square of the mass flow. However, more than pressure drop is affected by a change in mass flow. The value of h_0 changes as the (0.60 + 0.08 log d) power of mass flow, which affects the final value of the overall coefficient U_o, which affects the heating surface and consequently the length of tube required. Tube length also finally affects pressure drop. Therefore, not only is ΔP affected by G_2 but also by U_o and heating surface.

If the tube side calculated pressure drop had not been close to the allowable limit, this would also have required another trial at a mass flow rate G_1 either higher or lower than the value of 4 pps/sq ft in the trial calculation.

Note that the combustion products exit from the heat exchanger at 1084°F. This high temperature is, of course, a result of the limited heat recovery possible in heating the combustion air stream.

Obviously, a substantial quantity of usable heat is still present. It can be recovered in a heat recovery boiler, water heater, or organic heat transfer fluid heater, which we discuss in the following chapters. However, assume there is a use for process steam at 125 psig (353°F), and that the 1084°F gas is used to generate this steam. It is also assumed that a simple boiler without an economizer, can operate at a final stack temperature of 425°F, which is

72°F above saturated steam temperature. In such a system, an additional 7,117,200 Btu/h can be recovered, which will produce 7025 lb of steam per h at 125 psig.

Other forms of heat recovery can also be used after the recuperator, such as an organic heat transfer fluid heater, water heater, or higher pressure boiler, depending process needs. We discuss these forms of heat recovery in subsequent chapters.

Any additional heat recovery device used by the system after the air heater introduces another pressure drop. It should be known in advance that another heat recovery device will be used so that due allowance can be made in the overall pressure drop. Generally, heat recovery heaters and boilers are designed to a pressure drop of 4 to 6 in.

DIFFERENTIAL EXPANSION

The differential expansion between the tubes and the shell must be considered in any successful heat exchanger. This may be accomplished by any of the following means:

1. Expansion bellows section in the shell, with rigid tube sheets at both ends
2. Floating tube sheet at one end
3. Rigid tube sheets at both ends, with individual expansion bellows on each tube at the cold end

The use of a bellows section in the shell is not always practical, particularly on rectangular shells or shells lined internally with refractory material. The internal lining of the shell with castable refractory is quite common, particularly when the shell side gas is at high temperature (over 750°F), which is done to permit the use of carbon steel as the shell material.

A floating tube sheet at the cold end is a practical solution. The tube sheet may be attached to a flexible membrane or bellows-type convolution, or it can be designed as a packed gland expansion device. This design can be applied either to a rectangular or a circular tube sheet, although a circular tube sheet is usually better.

The use of individual expansion bellows at the cold end of each tube is probably the most expensive but also the surest method of taking care of differential expansion. It is virtually impossible to make all tubes in a heat exchanger operate at the same temperature. Tube temperature can vary with the actual temperature distribution across the face of the tube sheet and tube bundle and with the actually unequal flow through all tubes. Both methods 1 and 2 will accommodate differential expansion only as related to the actual average temperature of the shell. They cannot react to differential expansion tube-to-tube within the tube bundle. Only method 3 can react to individual tube temperature differences, and it should be used when high differentials exist and/or when tubes are very long. For short tube heat exchangers (15 ft and less), methods 1 and 2 can probably be used. For tubes over 15 ft with high differentials, method 3 should be used. Each case should be evaluated on its own merits.

See Fig. 32 for methods of handling differential expansion.

Figure 33 is a photograph of a typical rotary regeneration air-to-air heat exchanges.

Figure 34 illustrates a typical tube-type gas-to-air heat exchanges.

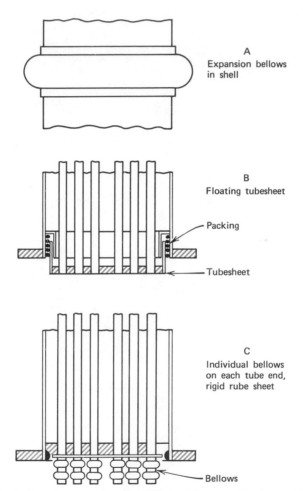

Fig. 32. Methods for compensating for differential expansion in straight tube heat exchangers.

Fig. 33. Rotary regenerative heat exchanger. Courtesy The American Schack Company, Inc.

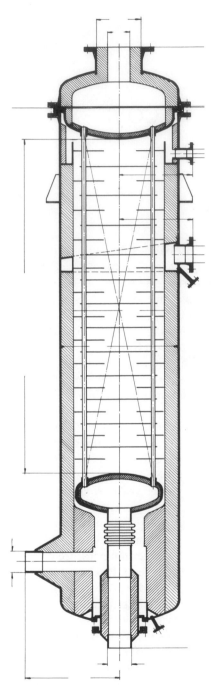

Fig. 34. Tube-type heat exchanger. Courtesy
The American Schack Company, Inc.

GAS-TO-WATER HEAT RECOVERY

USES AND ADVANTAGES

The most popular heat transfer medium is probably still steam, because it carries tremendous heat energy per unit weight, consisting of sensible and latent heat. However, pipelines are sized to carry certain *volumes* rather than weights of heat transfer fluid. The specific volume of steam is high. For example, 100,000 lb/h saturated steam at 100 psig can release 88 million Btu/h by condensation. If water is cooled from 450 to 300°F, 550,000 lb/h water would be needed to release the same 88 million Btu/h.

The volume of 100,000 lb saturated steam at 100 psig is 389,000 cu ft, requiring a 12-in. pipeline (and valves) for transport. The volume of 550,000 lb/h water at 450°F is 10,650 cu ft, requiring an 8-in. pipeline (and valves but no steam traps).

A high temperature water system has other advantages. It is a closed circuit; therefore no scale forming matter is introduced into the system continuously. Less scaling means less maintenance and longer life. Only the minimum in water treatment equipment is required.

Water systems can handle temporary peak loads easily. A steam system has to produce the actual steam demanded at any moment. The steam boiler must be sized for the peak load demand. High temperature water heaters are usually sized for about 80% of the peak load since they store enough sensible heat to handle temporary peak loads by larger temperature drop in the system than at normal load (thermal flywheel effect). It is not necessary to change the firing rate as suddenly and at such a wide turndown ratio as in steam generators. Smaller diameter pipe systems, minimum water treatment, and lower rated heater represent considerable savings over the steam system.

In the case of steam generators, economical considerations decide whether a fire-tube- or a water-tube-type boiler should be used. For lower capacities (and pressures) fire tube boilers are usually used, and for larger capacities, exclusively water-tube-type boilers. Fire-tube-type steam boilers are as trouble-free as water-tube-type boilers because evaporating steam on the shell side assures a film coefficient in the range of 4000 Btu/(h)(sq ft)(F°). This high coefficient keeps the metal temperature of the heated fire tubes and the unheated shell very close; therefore no harmful thermal stresses arise. If, however, water is not evaporated, the film coefficient is a function of the water velocity and is in the range of 2000 Btu/(h)(sq ft)(F°) at 0.30 fps water velocity. In the large diameter shell of a fire tube boiler, it is practically impossible to maintain a water velocity of this magnitude. This condition means that the metal temperature of the fire tubes is considerably higher than that of the shell, and this temperature difference creates harmful thermal stresses. For this reason, tube sheet failures are more common in fire tube high temperature water heaters than in boilers. Also for this reason, the temperature rise in a fire-tube-type high temperature water heater is limited to a maximum of 40°F. At higher temperatures, tube joint failures can be expected. In water tube heaters, the tube bends take care of any thermal stresses and expansion without harmful effect. Therefore, water tube units can be designed for any temperature rise. Comparing a water-tube-type high temperature water heater heater of 200°F temperature rise with a fire-tube-type unit of 20°F rise, ten times more water flow is required in a fire tube system of roughly ten times larger cross section at considerably greater expense. Therefore, water-tube-type high temperature water boilers are always more economical.

APPLICATIONS

Gas-to-water heat recovery is used when recoverable heat is most efficient for either high temperature hot water heating systems or in medium temperature hot water applications for absorption air conditioning. The driving heat source for an absorption system is usually low pressure steam. However, equipment is available to operate efficiently with water at medium temperatures. High temperature water systems operate at up to 400°F. Medium-range temperature systems operate between 230 and 280°F.

The source of hot gas for these applications is usually the exhaust from a gas turbine that may be used in a total energy system to drive an alternator or a mechanical refrigeration compressor to provide additional cooling. Gas turbine exhaust temperatures are in the range of 750 to 1000°F and, therefore, may be considered in the medium temperature range, requiring no special materials, alloy steels, and so on in the heat exchange sections.

In a total energy system used to provide all power, light, and heating and cooling for a hotel, school, shopping center, or hospital, a careful analysis of loads versus time must be made to determine the actual balance between electric power requirements and heat or air-conditioning requirements. Maximum heat recovery can, of course, be attained when gas turbine loads are at or near their maximum. If the maximum heating or air-conditioning load coincides with minimum or reduced electrical requirements, then the heat recovery units must either be supplementary-fired, or additional-fired water heaters must be added to make up the load. This condition is not economical because the gas turbine fuel rate is high under reduced loads; moreovers, this additional fuel must be burned either as supplementary firing of the heat recovery units or in stand-by fired water heaters.

One way to overcome this drawback is to use electrically heated water heaters that can be switched on to provide additional water heating under low base load conditions. This method provides the additional heating or cooling and allows the gas turbine to operate at a higher load and consequently better fuel rates. Naturally, fuel consumption of the gas turbine at reduced load plus the additional fuel consumption of the stand-by water heaters must be balanced against fuel consumption of the gas turbine only when operating to supply base load plus electrical water heating load.

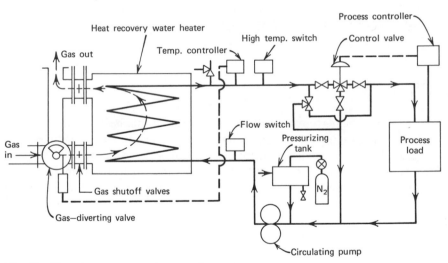

Fig. 35. Closed-system water heater schematic.

Heat recovery water heaters are operated in a closed pressurized loop with a circulating pump and pressurization equipment. See Fig. 35 for a typical schematic of a heat recovery water heater system.

KINDS OF GAS-WATER HEAT RECOVERY UNITS

The heating surfaces may be designed either as a rectangular grid bank made up of (1) sections of straight tubes set into headers at both ends, (2) sections of serpentine (hairpin-shaped) elements, (3) or circular sections or stacks of spiral (pancake) coils. Figure 36 shows a typical coil bank made up of spiral coils. Note that the ends of the individual spiral coils are attached to pipe headers, which are parallel to all coils. Inspection and clean out plugs may be installed in the headers opposite each tube to facilitate inspection and cleaning. The circular or spiral form inherently accommodates differential thermal expansion, which is definitely a plus for this design.

Fig. 36. Photo spiral coil bank. Courtesy Conseco, Inc.

Gas flow may be (1) from outside the coil inward radially toward the coil center, or (2) from the center radially toward the outside. Water flow is countercurrent to the gas flow, whichever gas flow path is selected. Generally, it is less complicated to flow the gas from outside the coil toward the inside. This path is used when gas temperature is 1000°F or less. When gas temperatures exceed 1000°F it is better to flow gas from the coil inside diameter to the outside, because this method avoids expensive casing lining required for higher temperatures.

An advantage of flowing gas from the coil outside diameter to the coil inside is that the longest turn (greatest heating surface) is exposed to the highest gas temperature, and since the average gas passage area (through flow area) reduces as the coil center is approached and the gas cools as it approaches the coil center, the mass flow [lb/(sec)(sq ft)] increases and the gas velocity remains high, thereby improving heat transfer throughout the coil.

If the waste gas contains considerable quantities of tube-fouling agents such as unburned hydrocarbons, tars, and the like, widely spaced bare tubes must be used, with adequate steam- or air-operated soot blowers strategically located to provide full coverage. Gas turbines that burn either light fuel oil or natural gas do not have such contaminants in their exhaust as a rule (except for oil-fired turbines, which may expel light soot during start up or load changes). Therefore, heat recovery units for such turbines may use finned tubes for heat transfer units.

USE OF EXTERNALLY FINNED TUBES

When the thermal resistance on the inside of a tube is much lower than that on the outside, as when steam condensing in a pipe is being used to heat air, or when water circulated in a pipe is heated by flowing gas over the pipe, externally finned surfaces, or extended heating surfaces, increase substantially the rate of heat transfer per unit length of pipe. For finned pipes in ordinary air heating practice, the temperature of the tip of the fin is nearly equal to that of the pipe.

For a given mass velocity of air, (1) the highest coefficients are for flow normal to cylindrical fins, (2) at intermediate values for flow parallel to discontinuous strip fins, and (3) the lowest coefficients for flow between parallel continuous fins. Serrated or segmented fins have some important advantages over solid fins since fin serrations cause turbulence in the gas passing over the outside of the tube. This turbulence results in maximized heat transfer coefficients. Independent research and field performance clearly demonstrate the superior heat transfer of the serrated over the continuous-type fin.

A ratio increase of over 2 to 1 in the heat transfer coefficient is obtained by the use of a multiplicity of $\frac{1}{8}$-in.- wide strip fins, suitably spaced, instead of continuous fins. A still greater increase is obtained by yet smaller strip widths. The heat transfer coefficient increases inversely as the square root, approximately, of fin width or perimeter. Fin thickness has but slight effect on the heat transfer coefficient of strip fins (when this coefficient is based on the *average* fin temperature, not the temperature at the base of the fin). Halving the thickness of fin width (4 instead of 8 percent) decreases the coefficient by only 4 to 12%. Thorough testing has shown that a segmented or cut fin surface increases substantially the values of the external film coefficient, with only a small drop in pressure.

Cleaning external fins can also pose a problem. The serrated fin, however, is easier to clean than the continuous fin, because the space between segments permits lateral flow of liquid, steam, or gas during the operation. This keeps buildups between the fins from accumulating rapidly and permits more effective steam cleaning. Figure 37 illustrates a typical segmented finned tube.

Fig. 37. Serrated-finned tube. Courtesy Escoa Fintube Corporation.

WATER-SIDE PRESSURE DROP

In heat recovery water heaters, water pressure drop should be held to less than 10 psi. Gas pressure drop should be held to less than 6 in. wg except in extreme cases. Water pressure drop is significant also because a reasonable pressure drop (6 to 10 psi) should be maintained for good distribution of water among the several parallel coils or tubes. In designs that do not permit such a pressure drop in the tubes, inlet restrictors or orifices should be used instead. Orifices are not desirable, however, because they tend to become clogged. A resistance consisting of several feet of small-diameter tubing is preferred because its i.d. can be greater than the orifice, permitting rather large particles to pass through. The safest method is to take the pressure drop in the heat transfer tube itself.

HEAT TRANSFER

Internal (liquid-side) heat transfer is improved by the higher pressure drop, indicating a higher mass velocity. External (gas-side) heat transfer is also improved by the higher gas pressure drop, indicating a higher gas mass velocity. This varies approximately as $G^{0.6}$, where G = the gas mass velocity. Therefore, these factors must be considered in a well-balanced design.

Outside Tubes

For both liquids and gases flowing normal to banks of unbaffled staggered tubes, for DG_{max}/μ_f exceeding 2000, the following equation can be written:

$$\frac{h_m D_o}{k_f} = 0.33 \left[\frac{C_p \, \mu_f}{k_f} \right]^{1/3} \left[\frac{D_o G_{max}}{\mu_f} \right]^{0.6} \tag{32}$$

For in-line tubes, the constant is reduced from 0.33 to 0.26. Subscript "f" refers to fluid property.

A simplified dimensional equation for gases flowing normal to staggered tubes is

$$h = 0.133 \, C_p G^{0.6}/D^{0.4} \tag{33}$$

where D is in feet and $G/(\text{lb})(\text{h})(\text{sq ft})$. However, even Equation 33 gives more optimistic values than Equation 12, which is plotted on Fig. 18 and will be used throughout this work.

Inside Tubes

Coefficients of heat transfer for water in turbulent flow inside tubes or pipes is taken from Figs. 40 and 41.

When designing a gas-water heat recovery system, one must bear in mind practical considerations of continued operation, convenient service and maintenance, and easy replacement of componenets. Figure 18 illustrates the effect of tube diameter on the coefficient of heat transfer. Obviously, the smaller the tube, the greater the coefficient, with consequently less heating surface required. Therefore, theoretically, the tubes should have small diameters. Practically, however, even closed-system units such as a closed water loop always have some pump seal leakage, draining losses, and so on that require makeup water. Thus, some deposition of solids will occur, even over a long period of time. Extremely small tubes are more vulnerable to this than larger tubes because the tube surface exposed to a given rate of water flow is so much less in smaller tubes. For this reason, tube size for closed-system water heaters should not be smaller than $\frac{3}{4}$-in. o.d., with 1-in. o.d. preferable.

EXAMPLE 81

Fin Selection

Likewise, on the gas side, the hot-gas-condition entrained solids and the like should be considered in selecting the external fin configuration. Natural-gas-fueled gas turbines produce a very clean exhaust; consequently high, closely spaced fins may be used, up to a fin spacing of 10 fins per in. Fin height, while adding to total surface, reduces fin efficiency because of the longer thermal path. Higher fins also increase through flow area or gas passage area. All these factors must be balanced in a design for most effective use of tubes. This involves an evaluation of water-side velocities and pressure drop, gas-side mass flow and pressure drop, and average fin temperature.

Light-oil-fueled gas turbines, using fuels such as No. 2, JP5 or kerosene, produce relatively clean exhaust under steady load. However, under rapid load changes or at start up, some carbon or soot may be deposited. For this reason, fin spacing should not be closer than 8 fins per in., and the preferable fin height should not exceed $\frac{5}{8}$ in. for easy cleaning.

Hot waste gas from other sources burning heavier fuels, municipal wastes, or sewage sludge and at higher temperatures must be evaluated separately. Heat recovery units operating on heavy fuel oils should have several unfinned rows ahead of finned sections, and finned sections should have fins spaced not closer than 6 fins per in. and not higher than $\frac{5}{8}$ in. All tubes of heat recovery units on municipal waste incinerators should be bare, as fins will clog very rapidly. Heat recovery units on sewage sludge incinerators should have several unfinned rows and then fins not closer than 6 fins per in. and preferably 4 fins per in., and not over $\frac{5}{8}$ in. high. Properly located soot blowers should always be included to cover all surfaces.

Gas temperatures over 1200°F may result in substantially high fin tip temperatures. This must be determined by methods to be demonstrated later, with fin thickness and material to satisfy conditions. Usually, low carbon steel tubes may be used to ASME Specification SA178, Grade A, with either carbon steel or chrome steel fins, depending on temperature. Stainless steel tubes and fins are used only under extremely corrosive gas conditions.

EXAMPLE

The best way to demonstrate the aforementioned points is through a practical example. A gas turbine driving an alternator is used as the heat source. The turbine exhaust will go to a heat recovery water heater. The heated water is used in a closed loop to drive an absorption chiller for air conditioning.

The gas turbine selected is a Solar Saturn 1200-hp unit. The following are full load performance data for this turbine at the site altitude. (Refer to

Solar Saturn turbine performance curves in the Appendix Fig. A-f-3

Ambient temperature—80°F
Altitude—3000 ft
From Fig. 3, altitude correction factor—0.902
Maximum sea level power (continuous)—812 kW
Sea level gas flow—48,600 lb/h
Power at altitude—732 kW
Gas flow at altitude—43,837 lb/h
Exhaust gas temperature—860°F
Gas turbine fuel—natural gas

Hot water requirements are as follows:

Water flow rate—290 gpm
Water return temperature—240°F
Water supply temperature—280°F
Water pressure drop (maximum)—10 psi
Gas pressure drop (maximum)—6 in./wc

$$\text{Required heat recovery} = (290)(500)(280-240)$$
$$= 5,800,000 \text{ Btu/h}$$

In determining the exhaust gas temperature drop, assume a bypass leakage loss of 4% and radiation losses of 1%. The bypass leakage loss is the gas loss through the bypass valve in the fully closed position. This is taken because under practical conditions it is difficult to manufacture a modulating valve that is also 100% tight in the closed bypass position.

$$\text{Gas } \Delta T = \frac{5,800,000}{(43,837)(0.95)(0.256)} = 544°F$$

where 0.95 is the leakage factor and 0.256 is the average specific heat of the gas from Fig. 2.

$$\text{Heater stack temperature} = 860 - 544 = 316°F.$$

The stack temperature of 316°F is the result of the coil exhaust temperature and the higher leakage loss gas temperature. This temperature may be determined by the following:

$$(43,837)(0.95)(T_2) + (43,837)(0.05)(860) = (43,837)(316)$$

$T_2 = 287°F$, which is the coil exhaust temperature and the terminal temperature used in calculating Lm ΔT for the coil bank.

EXAMPLE   83

The conditions for the heat recovery water heater are then as follows:

Heat recovery—5,800,000 Btu/h
Water rate—290 gpm
Water temperature in—240°F
Water temperature out—280°F
Gas temperature in—860°F
Gas temperature out—287°F
Net gas flow—43,837 lb/h = 12.2 pps
Water pressure drop (maximum)—10 psi
Gas pressure drop (maximum)—6 in./wc
Bypass leakage loss (maximum)—4%
Radiation loss (maximum)—1%
Stack temperature—316°F
Lm ΔT—212°F

ASME Requirement

The selection of tube size is somewhat arbitrary. Since the proposed unit is reasonably small, operating in a closed system, and with a natural-gas-fired turbine, 1.25-in. o.d. tubing can be used. The maximum water temperature is 280°F. At this temperature the coil must be pressurized to at least 75 psig to prevent nucleate boiling, and the design pressure will be 90 psig. Tube wall thickness must be determined by the formulas set forth by the ASME Power Boiler Code, Section I, Paragraph 27.2.1.

$$t = \frac{PD}{2S + p} + 0.005D \qquad (34)$$

where t = tube wall thickness, min
   P = design pressure, 90 psig
   D = tube o.d. = 1.25 in.
   S = 11,500 psi for SA178 Grade A at 700°F (from ASME Table PG23.1)
   t = 0.011 in.

Obviously, this is an impractical tube wall thickness for both fabricating and welding. The minimum tube wall that should be used here is 0.083 in. (14 gauge). The average i.d. for 1.25-in. 14 gauge minimum wall tube is 1.072 in.

A reasonable water velocity for this kind of application is 6 fps. If this selection results in a water pressure drop in excess of 10 psi, then another trial must be made. Six fps velocity in an inside diameter of 1.072 in. requires a flow rate of 16.9 gpm per tube. For the 290 gpm, 17.2 coils or tube circuits are required. For a coil bank of 17 parallel coils, the flow rate per tube is 16.76 gpm, and the average water velocity is 5.95 fps.

External Fin Configuration

The next step is to determine the external fin configuration and total heating surface. Because the fuel is natural gas, 10 fins per in. are used. Fin height is governed by the required through flow area and the maximum gas pressure drop allowance. Figure 38 is a curve of gas pressure drop versus mass flow for coiled tubes finned 10 fins per in. This curve was developed from actual tests on heat recovery units designed for industrial installations. The circular coil is selected for the example because it is ideally suited to heat recovery conditions of the magnitude in this example. Again, a trial must be made.

Gas Pressure Drop

Gas pressure drop through the coil is not the only pressure drop to be considered. There is also gas entry loss, bypass valve loss (particularly if there is a 90° turn at the inlet), coil pressure drop, contraction at the coil center, expansion at the coil periphery, and contraction at the gas exit. Figure 39 illustrates these conditions at varying rates of mass flow. The cross section at each point must be calculated to find the mass flow. For a total gas pressure drop of 6 in., start with a coil pressure drop of 2 in. From Fig. 38 the gas mass flow for a 6 turn- (pass) coil at 2 in. pressure drop is 1.375 pps per sq ft.

Total gas flow = 43,837 lb/h = 12.2 pps
Gas passage area = 12.2/1.375 = 8.873 sq ft
Gas passage area per coil = 8.873/17 = 0.522 sq ft (average)

The 0.522 sq ft is the gas passage area of the average turn in a multiple-turn coil.

At this point, assumptions about fin height and thickness must be made. For a trial, use a fin height of 0.375 in., a thickness of 0.025 in., and a fin tip-to-tip spacing averaging 0.030 in. This configuration produces a net gas passage or through flow area of 0.051 sq ft per ft of tube length. The total external heating surface as taken from the Escoa Fintube Corporation manual is 3.17 sq ft per ft.

The length of the average turn is then 0.522/0.051 = 10.23 ft, and the total length of a six-turn coil is 61.4 ft. The total heating surface for one coil is (61.4)(3.17) = 195 sq ft. The heating surface of a bank of 17 coils is 3315 sq ft.

Next determine the required heating surface as a check on the preceding operation by the use of the methods developed.

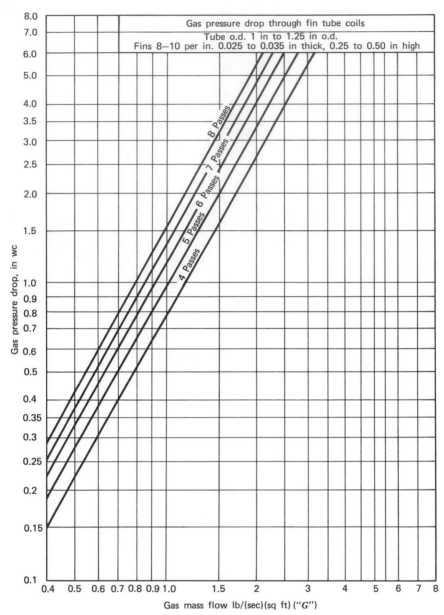

Fig. 38. Finned-coil, gas pressure drop ΔPg versus mass flow G.

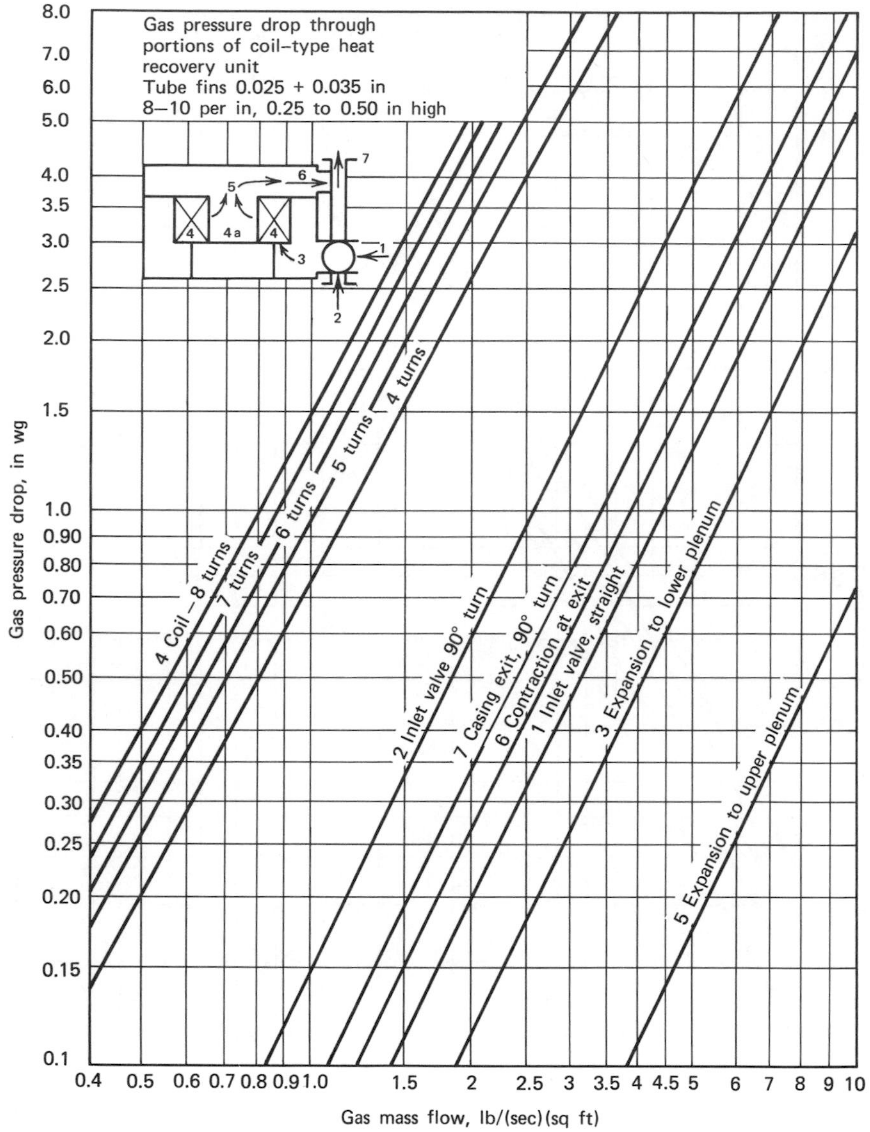
Gas pressure drop through portions of coil-type heat recovery unit
Tube fins 0.025 + 0.035 in
8–10 per in, 0.25 to 0.50 in high
Gas pressure drop, in wg
8.0
7.0
6.0
5.0
4.0
3.5
3.0
2.5
2.0
1.5
1.0
0.90
0.80
0.70
0.60
0.50
0.40
0.35
0.30
0.25
0.20
0.15
0.1
4 Coil–8 turns
7 turns
6 turns
5 turns
4 turns
2 Inlet valve 90° turn
7 Casing exit, 90° turn
6 Contraction at exit
1 Inlet valve, straight
3 Expansion to lower plenum
5 Expansion to upper plenum
0.4
0.5
0.6
0.7
0.8
0.9
1.0
1.5
2
2.5
3
3.5
4
4.5
5
6
7
8
9
10
Gas mass flow, lb/(sec)(sq ft)

Fig. 39. ΔP through portions of a heat recovery unit.

EXAMPLE 87

Gas side:
Gas flow rate—12.2 pps
Average gas passage area—8.873 sq ft
Gas mass flow G = 12.2/8.873—1.375 pps/sq ft

From Fig. 18, $\dfrac{G^n}{D^{.53}}$ for 1.25-in. tube = 1.15

Average gas temperature T—927°R
From Fig. 19, for T = 927, $T^{1/3}$ = 9.73

$$h_o = (G^n)(T^{1/3}) = (1.15)(9.73) = 11.19 \text{ Btu/(h)(sq ft)(F°)}$$

Water side:
Tube i.d.—1.072 in.
Water rate per tube—16.76 gpm
Water velocity—5.95 fps
Average water temperature—260°F
From Fig. 40, for V = 5.95, h_i = 1150
From Fig. 41, for D = 1.072, F_d = 0.99, and for T_{av} = 260°F, F_t = 1.60

$$h_i = (1150)(0.99)(1.60) = 1822 \text{ Btu/(h)(sq ft)(F°)}$$

Tube Wall:
Minimum tube wall—0.083 in.
Average tube wall—0.089 in.
Material—low carbon steel
Thermal conductivity—25 Btu/(F°)(ft)(h)

$$h_w = \frac{25}{0.089/12} = 3371 \text{ Btu/(h)(sq ft)(F°)}$$

Fouling:
For relatively clean gas, R_{fo} = 0.001
For closed water system, R_{fi} = 0.001

Fin efficiency:
Gas-side coefficient—11.21
Fin thickness—0.025 in.
Fin height—0.375 in.
Fin material—low carbon steel
From Fig. 42, for the preceding fin conditions, fin efficiency E = 0.85
Ratio of outside surface to inside surface:

$$\frac{A_o}{A_i} = \frac{3.17}{0.281} = 11.28$$

$$R_o = \frac{1}{(11.19)(0.85)} = 0.105160$$

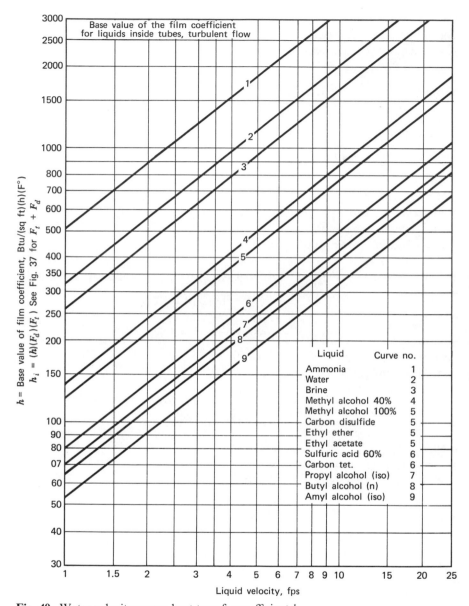

The chart shows:

Vertical axis label: h = Base value of film coefficient, Btu/(sq ft)(h)(F°)

$h_i = (h)(F_d)(F_t)$ See Fig. 37 for $F_t + F_d$

Title inside chart: Base value of the film coefficient for liquids inside tubes, turbulent flow

Liquid	Curve no.
Ammonia	1
Water	2
Brine	3
Methyl alcohol 40%	4
Methyl alcohol 100%	5
Carbon disulfide	5
Ethyl ether	5
Ethyl acetate	5
Sulfuric acid 60%	6
Carbon tet.	6
Propyl alcohol (iso)	7
Butyl alcohol (n)	8
Amyl alcohol (iso)	9

Horizontal axis label: Liquid velocity, fps

Fig. 40. Water velocity versus heat transfer coefficient h_o.

EXAMPLE   89

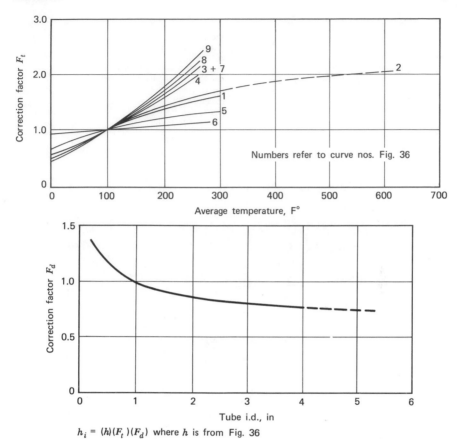

$h_i = (h)(F_t)(F_d)$ where h is from Fig. 36

Fig. 41. Temperature and diameter correction factors for Fig. 36.

$$R_i = \frac{11.28}{1822} = 0.006191$$

$$R_w = \frac{1}{3371} = 0.000297$$

$$R_{fo} = 0.001000$$
$$R_{fi} = 0.001000$$
$$R_t = 0.11365$$

$$U_o = \frac{1}{R_t} = 8.80 \text{ Btu/(h)(sq ft)(F}°)$$

Lm $\Delta T = 212°$F

Required heating surface $= \dfrac{5,800,000}{(8.80)(212)} = 3109$ sq ft

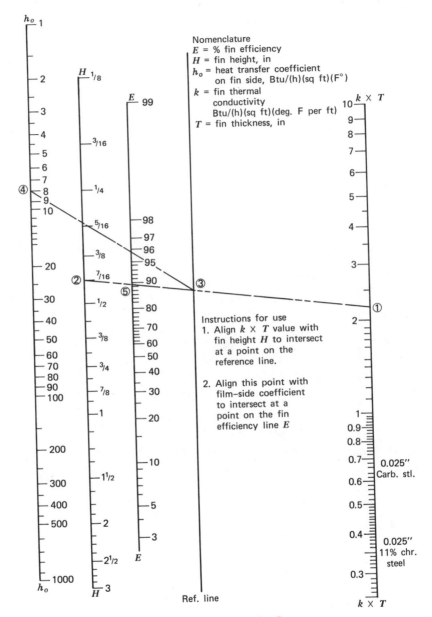

Fig. 42. Fin efficiency. Courtesy, Escoa Fintube Corp.

EXAMPLE   91

The coil determined previously for gas pressure drop has 3315 sq ft. Therefore, the check figure of 3102 is sufficiently close, allowing approximately 7% in excess of calculated heating surface requirement. If a closer approach is desired, then another trial must be made.

Pressure Drop—Gas-Side Check

Pressure drop must now be checked to determine whether the coil described above also meets these requirements. Assume that the circular heat transfer coil is installed within an insulated casing with gas flowing into the casing, entering the coil on the outside diameter and leaving through the coil inside diameter and then out to the stack. Refer to the schematic diagram in Fig. 39. The inlets at 1 and 2 represent the entrances to a gas bypass control valve, which is controlled by water outlet temperature to admit or bypass portions of the hot gas as required to meet load conditions.

Gas conditions through the coil are as follows:

Mass flow—1.375 pps/sq ft
Gas temperature in—860°F
Gas temperature out—287°F
Average gas temperature—573°F
Number of gas passes—6
Gas flow through coil—12.2 pps
Gas i.d.—24 in.
Coil l.d.—26.375 in.
Gas o.d.—26 in.
Target gas pressure drop—6 in./wg

Assume that the hot gas enters the bypass valve at point 1 in Fig. 39, going directly into the casing at point 3, through the coil radially from 3 through points 4 to 5, and then out by way of points 6 and 7. There will be considerable turbulence in the coil l.d. region, because the gas is coming out radially toward the center before proceeding vertically out of the coil center. Designate this zone as 4a and consider it as 90° mitre elbow for pressure drop purposes, using Curve No. 7 for gas at exit temperature in a 90° mitre condition.

Point 1 mass flow, 3.88 pps/(sq ft)(ΔP_1)   = 0.750
Point 3 mass flow, 3.88 pps/(sq ft)(ΔP_2)   = 0.440
Point 4 mass flow, 1.375 pps/(sq ft)(ΔP_4)   = 1.940
Point 4a mass flow, 3.22 pps/(sq ft)(ΔP_{4a}) = 0.900
Point 5 mass flow, 3.22 pps/(sq ft)(ΔP_5)   = 0.080
Point 6 mass flow, 3.30 pps/(sq ft)(ΔP_6)   = 0.780
Point 7 mass flow, 3.30 pps/(sq ft)(ΔP_7)   = 0.980

   Total gas pressure drop   5.870

The limit was set at 6 in./wg. Therefore, the design value is sufficiently close for all practical purposes. The preceding method has been tested on actual production equipment utilizing the circular coil concept. The coil pressure drop data are for an average gas temperature of 570 to 650°F. If other temperatures are encountered, then the curve data must be corrected accordingly. For example, assume that the average gas temperature in the coil is 800°F; then the correction factor is $(800 + 460)/(605 + 460) = 1.183$, where 605 is the average temperature F°, $(570 + 640)/2$, at which the curves were developed.

If the equipment uses a rectangular coil section made up of return-bend sections instead of the circular coil design, the coil's net average gas through flow area must be determined so that mass flow area must be determined so that mass flow rate can be calculated, then proceed in the same way, using appropriate curves, depending on the gas path through the system.

Pressure Drop—Water-Side Check

The pressure drop through the water side must now be checked to determine whether the aforementioned coil also meets these requirements. Assume that the 17 coils are welded into 4-in. pipe headers at both ends. The cooler water enters the header connected to the inside of the coil ends and the hot water leaves the coil through the header connected to the outside of the coil ends.

Water flow rate—290 gpm
Water temperature entering—240°F
Water temperature leaving—280°F
Average water temperature—260°F
Number of tube coils—17
Tube I.d.—1.072 in.
Flow per tube coil—17.06 gpm
Effective heat transfer coil length—61.4 ft
Actual length header-to-header—64.4 ft
Water velocity—6.06 fps
Maximum pressure drop—10 psi

Since the coil is circular, some allowance must be made for the effect of coiling the tube on pressure drop. Figure 27 can be used, but it must be extrapolated to the R/D ratio of 18.2 that exists in subject coil. Using the extrapolated value of 25 tube diameters, and allowing for six 360° turns, the additional tube length for curvature is $(25)(1.65)(1.072)/12 = 3.69$ ft. The total equivalent tube length is then $64.4 + 3.69 = 68.09$ ft.

From Fig. 43, water pressure drop in coiled tubes, for a velocity of 6.06 fps and tube I.d. of 1.072 in., $\Delta P/N = 0.10$ psi per ft. Coil pressure drop is $(68.09)(0.10) = 6.809$ psi.

EXAMPLE   93

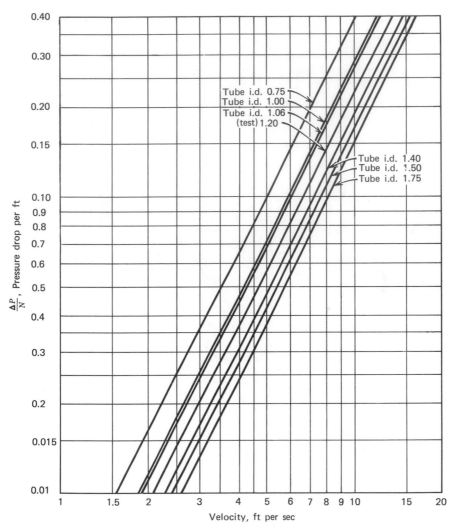

Fig. 43. Water pressure drop in coiled tubes. Average radius, 22 in.

Figure 44 is for pressure drop resulting from a sudden enlargement, such as a smaller diameter tube discharging into a header.

For a 1.072-in. l.d. tube discharing into a 4-in. pipe header, $D/d = 3.76$. From Fig. 40 for $V = 6.06$, head loss $= 0.45$ ft $= 0.016$ psi.

Figure 45 is for pressure drop resulting from a sudden contraction, such as a flow of water from a header into a smaller pipe.

For a 1.072-in. l.d. tube and a 4-in. header, head loss $= 0.29$ ft $= 0.01$ psi. Total pressure drop is then 6.835 psi, well below the upper limit of 10 psi.

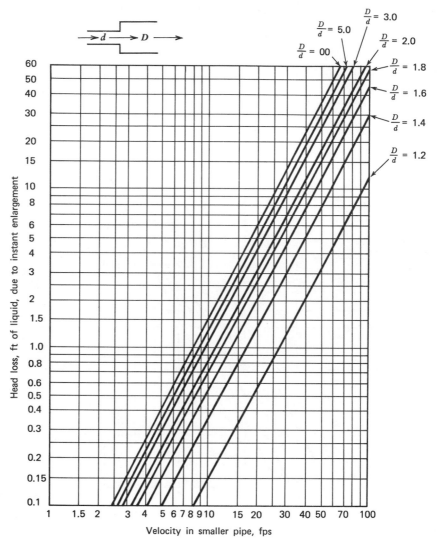

Fig. 44. Loss of head in feet of any liquid due to sudden enlargement.

The heating surface coil as calculated above is within the established design criteria; if properly manufactured, it will operate in accordance with requirements.

CONTROL

The normal control method is by means of a gas bypass valve pneumatically or electrically operated and regulated by a temperature controller, regulating

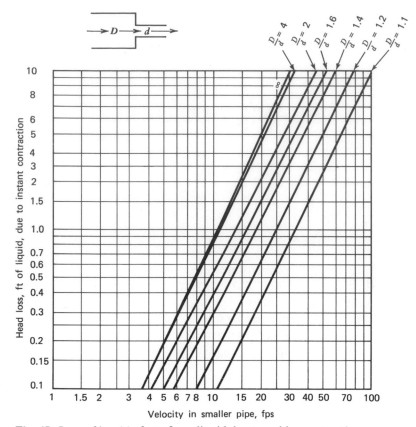

Fig. 45. Loss of head in feet of any liquid due to sudden contraction.

water outlet temperature. This must be a modulating-type control sufficiently responsive to achieve accuracy within $\pm 2°F$ when controlling at
280°F. Another, not so common, method is to place a controllable heat sink
into the hot water system to divert a portion of the hot water supply from
the heat recovery unit into a heat exchanger, thereby rejecting surplus heat
to cooling water.

In addition to the control under discussion, a heat recovery water heater
must also be equipped with an excess water temperature switch and a flow
switch connected into the system to fully bypass hot gas in the event of
excess water temperature or inadequate water flow.

A heat recovery water heater must be designed and constructed in accordance with applicable rules of the ASME Boiler Code Section I and must be
equipped with safety relief valves as required by the code.

GAS-TO-ORGANIC FLUID HEAT RECOVERY

USES AND APPLICATIONS

The use of organic heat transfer fluids in plant process heating applications has grown rapidly over the past several years and continues to have many varied applications. The greatest advantage of using these fluids lies in their high temperature–low pressure characteristics. A steam or hot water system operating over 212°F must be pressurized to the desired temperature. For example, a steam system operating at 500°F requires a pressure of 680.8 psia. In contrast, an organic heat transfer fluid system can operate at 500°F at atmospheric pressure.

The major economies in using organic heat transfer fluids over steam or other vapor systems is the lower cost of installation and operation. Unpressurized or low pressure systems can significantly reduce capital cost. The savings in eliminating the installation of large-diameter vapor piping, flash drums, safety valves, and pressure control devices can amount to 25 to 50% of system costs. A second source of savings is in operation. Nonpressurized systems require minimum maintenance, eliminate condensation heat loss in pipelines, and usually do not require licensed operating personnel.

Some major advantages of an organic fluid system over oil are (1) greater versatility in operating temperature, (2) greater chemical stability, (3) reduced makeup, and (4) closer control of and faster response to temperature change as required by the process. Temperature control accuracy of $\pm 2°F$ can be easily achieved.

The number of organic heat transfer fluid applications is practically limitless. Some of these are listed below:

Petroleum refining
Brine liquors concentration and distillation
Tanker and barge heating

Synthetic fiber production
Metal treating: annealing, stress relieving
Rubber, plastics, and paper processing heat
Alkyd paints and resin kettle heating
Chemical process heating
Waste heat recovery systems
Offshore oil and gas platform heating
Space heating for large buildings
Bituminous materials heating
Converters: calendering and distillation

Several different organic heat transfer fluids are available. The fluid selected for a particular system depends on the temperature requirements of its operation. In any range, fluids are ideally suited for systems that must deliver uniform heat (up to 800°F), precise temperature control, and quick response to heating or cooling demand.

POINTS TO CHECK IN SYSTEM DESIGN

While there is nothing very complicated about a fluid heat transfer system, certain design parameters must be observed to assure trouble-free service, efficiency, and a return of all benefits of heating or cooling with a non-pressurized liquid.

The basic engineering checkpoints of designing a system are listed as follows:

1. **The Heater**

    a. To insure nondegrading delivery of heat from the fuel
    b. To be designed so that maximum allowable film temperature for the particular fluid is not exceeded at any point in the heater

2. **The Pumps and Piping**

    a. To insure proper handling of the fluid and to maintain the correct flux at user stations and adequate system circulation
    b. To assure that the circulating pump is selected for the proposed operating temperature (water-cooled seal and bearings if necessary)

3. **Materials and Construction**

    a. To insure compatibility and proper layout and design

With sufficient heater capacity, good control of flow at user stations, and compatible materials, a well-designed system gives reliable, efficient, and precise delivery of heat. Proper selection of fluid and a clean moisture-free system insures minimum maintenance and trouble-free operation.

Liquid heat transfer is basically heating an adequate quantity of fluid to a practical working temperature, circulating the fluid to heat-accepting user stations at a rate to raise or maintain the temperature of the heat user as required by the process. Therefore, in designing a system, the following questions should be answered:

1. Is the process a continuous or batch operation?
2. What is the heat demand and temperature requirement of each heat user? the cooling requirement?
3. How much heat transfer area is practical, or available, at user stations, and what coefficients of heat transfer can be expected from the product at user surfaces?
4. Will there be more than one heat user on stream at the same time? at the same temperature? at different temperatures?
5. What heat losses are expected in the piping layout of the system?
6. Can existing heating requirements be served by incorporations into the new system to eliminate older equipment or other types of heat delivery?
7. Are expanded heating requirements probable?

The basic starting point is to estimate the heat balances and determine the total heat demand of the system. This will guide the selection of heater capacity, reveal fluid operating temperature limits, and indicate the type of fluid the system requires.

Only fluids that are chemically and thermally stable should be selected. Such fluids resist chemical and physical changes at all temperatures below their maximum bulk temperature operating limits. These maximum temperatures are stated for each fluid by their suppliers. Unsubjected to contamination or heat stress beyond their limits, these fluids give long service without significant physical or chemical change. Test the fluid periodically to determine its condition as part of a good system maintenance.

The heat transfer fluid is used as a liquid-phase heating medium, so transfer of heat is by sensible rather than latent mode as with a condensing vapor. Heated liquid circulated at the higher velocities over a heat-using surface can be a more efficient and readily controlled method of heat transfer than pressure-controlled condensing vapor systems. As with any heat transfer fluid, system design must consider the relationship between liquid velocity and heat load from the heat source. Liquid film in immediate contact with the heat donor surface (whether the heater fuel is electricity, gas, oil, or a waste heat source) is subjected to higher temperatures than the bulk temperature of the balance of the fluid. The velocity of the circulating fluid is the critical determinant of how long the film should be subjected to the higher temperature. The differential between the fluid bulk temperature and the film temperature depends on the fluid velocity over the heater surface and the physical properties of the fluid used. Fluids having high specific heat and thermal conductivity should be selected wherever possible for maximum efficiency.

In selecting both the heater and the fluid, *maximum* rather than average film temperature should be considered and a reasonable margin allowed to enable the fluid to give the longest service. Each application for heat transfer fluids has its own engineering and cost requirements. Each heat transfer fluid, in turn, has its particular properties. The universal fluid does not exist. Therefore, in selecting a heat transfer fluid for a given application, the design engineer must balance his requirements against the performance that the fluid's properties can provide.

TYPES OF FLUIDS

The many different available fluids allow the system designer to choose the best fluid for his particular application. Fluids are broken down into the following general classifications:

1. Polyalkylene glycols have good thermal stability up to 500°F. They are easy to pump because of good viscosity—temperature and lubricating properties. They do not boil. Thus, pressurized systems are not required. Most types are noncorrosive, water soluble, and provide efficient heat transfer.
2. Glycols have an effective temperature range of -50 to 350°F. They are nonlubricating. They are used in water solutions in air-conditioning equipment, gas compressor engines, snow melting systems, skating rinks, and other closed heating and cooling systems. Specifically inhibited grades of ethylene, diethylene, triethylene, and propylene glycols are available.
3. Fluorocarbons are primary refrigerants in compressors for air conditioning and refrigeration. Some fluorocarbons freeze at -168°F and boil at 74.8°F.
4. Water is an excellent and the most widely used heat transfer medium, with high specific heat and heat of vaporization. However, pressure-temperature limitations and corrosion can be problems.
5. Mineral oils, low in cost and noncorrosive, are used as heat transfer fluids at temperatures from -15 to 600°F. However, they have low heat capacity and are susceptible to thermal cracking, which produces both volatiles and coke or sludge residues. Lower viscosity grades are easily

pumped. Higher viscosity oils, frequently used at higher temperatures, are difficult to pump at start up.

6. Silicate fluids cover a wide range of operating temperatures. Two available general types are alkyl and aryl silicates. Alkyl silicates such as inhibited tetra (2-ethylhexyl) orthosilicate, as supplied by Union Carbide, are useful over the temperature range of -40 to $600°F$. The alkyl silicates are useful for both heating and cooling, frequently using common equipment. The tetra-aryl silicates find application at 25 to $650°F$ and provide a degree of fire resistance. Both types are expensive and require closed systems to reduce oxidation or to eliminate moisture contamination that can cause hydrolysis to a gelatinous state.

7. Inorganic salt mixtures are the only commercially available products usable at temperatures of up to $1000°F$. Molten alkali metal systems are receiving serious consideration for this type of service. Since salt mixtures solidify below $300°F$, they must be kept in continuous operation and drained after shutdown. Contamination of salt mixtures changes the composition from a true eutectic, with a corresponding increase in melting point.

8. Diphenyl-diphenyl oxide eutectic mixture is used over a temperature range of 54 to $750°F$ and boils at $496°F$. A closed pressure system is required. The specific heat of the fluid is high. Since the fluid freezes at $54°F$, steam tracing of lines may be required. Extreme care must be employed with seals, packings, and fittings because this product has a marked tendency to leak, creating problems of fluid loss, odor, toxicity, and clean up. A well-known brand of this type of fluid is Dowtherm A. Dowtherm A may be used either in its liquid or vapor phase as a heat transfer medium.

In a process where heat uniformity is required, condensing vapor offers precision temperature control. An equivalent liquid system would have to be operated at extreme flow rates to maintain the same precision. Vapor-phase systems are also advantageous in installations where it is difficult to control liquid flow pattern and velocity, such as in jacketed kettles.

In systems where the temperature gradient within the product cannot exceed certain limits, low Δt within the heat transferring media must be maintained. This can be better accomplished with a condensing vapor than with liquid at high mass flow rates.

In liquid-phase systems, the liquid film coefficient increases with the velocity of the liquid and is adjustable over a wide range. In vapor-phase systems, the condensing film coefficient is fixed by temperature difference, tube size, and tube arrangement.

9. Chlorinated biphenyls are fire-resistant fluids with a useful range of 50 to $600°F$. At high temperatures, however, these fluids may liberate hy-

drogen chloride. A closed corrosion-resistant system is required. Chlorinated biphenyls have a low specific heat.

Physical properties of some of the more commonly used heat transfer
fluids are contained in Table 10. For more detailed data on a particular fluid,
consult the manufacturer's technical data.

THE SYSTEM

The Heater

The heater is the most critical component in designing a heat transfer fluid
system. With the proper balance of heating capacity, temperatures, and fluid
velocity, the service life of the fluid is maximized.

There are two basic types of fluid heaters: the liquid-tube type and the
fire-tube type. Either type may be fired or heated by a hot gas source such as
a gas turbine or an incinerator. The latter is the case for the heat recovery
system we are discussing. In liquid-tube heaters, fluid is pumped through the
tubes as it is heated. In fire-tube heaters, fluid flows through the shell of the
heater surrounding the fired tubes. It is more practical to have controlled
velocities in the liquid-tube type; since the fluid flows through the shell of
a fire-tube type, velocities vary greatly throughout the vessel, with consequent
danger of "hot spots." The liquid-tube type will be used throughout this
chapter since it permits better film temperature control. Fluid velocities over
the heater surfaces should be relatively high, generally 4 to 12 fps, to help
avoid excessive film temperatures that may be detrimental to the heater
surfaces and the fluid. A method for calculating maximum fluid film temperature will be demonstrated in the design example in this chapter.

The Pump

Pumps used must have sufficient capacity and pressure head to circulate the
fluid at the rate required by the system. For large-capacity systems, the pump
should be of the centrifugal type. A number of standard brands of high
temperature pumps are available. Pump manufacturers usually specify water-
cooled mechanical seals and bearings for temperatures over 450°F. Mechanical seal pumps are preferable to stuffing box pumps. Pumps with a stuffing
box should be provided with at least five rings of packing with graphite-
impregnated metal foil around an asbestos core or laminar graphite rings
such as Grafoil (by Union Carbide).

When a new system is first put into operation, a slight leakage may occur
at the pump packing. If the pump is a stuffing box type, the gland should not
be tightened until the system has heated up to close to the temperature of
operation. Canned pumps can be very serviceable in a hot fluid system.

TABLE 10 Properties of Some Heat Transfer Fluids

Property	Mineral Oils				Diphenyl Diph. Oxide	Ethylene Glycol
	Humble Therm 500	Mobil Therm 600	Tidewater Avalon 90	Sunoco Circo XXX	Dowtherm A	Union Carbid
Chemical stability	Stable	Stable	Stable	Stable	Stable, but water can be dangerous	Stable
Oxidative	Up to 150°F w/air	Up to 150°F w/air	Up to 150°F w/air	Up to 150°F w/air	Stable	Low oxidation
Thermal stability	High resistance to decomposition. Deposits carbon	High resistance to decomposition. Deposits carbon	Slow decomposition starts at 575°F, more at 750, rapid 850	Similar to tidewater Avalon 90	No appreciable decomposition below 650°F. Steady decomp., 650+	Thermally stable to 400°F
Max. Temp.						
Film	680°F	625°F	620°F	590°F	805°F	400°F
Bulk	615°F	560°F	555°F	525°F	700°F	325°F
Safety						
Fire Res.	None-burns	None-burns	None-burns	None-burns	None-burns	None-burns
Explosive	Negligible	Negligible	Negligible	Negligible	Explosive mists possible	None
Density, lb/g						
70°F	7.1	7.8	7.4	7.8	8.82	9.30
300	6.5	7.3	6.7	7.3	7.98	
600	5.5	6.4	5.9	6.5	6.60	Not used
700	Not used	Not used	Not used	Not used	6.03	
Spec. heat						
70°F	.48	.38	.43	.43	.379	.625
300	.58	.485	.55	.55	.458	.710
600	.72	.624	.70	.70	.560	Not used
700	Not used	Not used	Not used	Not used	.591	

TABLE 10   Properties of Some Heat Transfer Fluids (*continued*)

	Mineral Oils				Diphenyl Diph. Oxide	Ethylene Glycol
Property	Humble Therm 500	Mobil Therm 600	Tidewater Avalon 90	Sunoco Circo XXX	Dowtherm A	Union Carbide
Therm. cond.	Btu/(sq ft)(h)(F°)(ft)					
70°F	.078	.070		.070	.081	.167
300	.072	.065		.066	.072	
600	.065	.059		.060	.061	Not used
700	Not used	Not used	Not used	Not used	.057	
Pour point	+15°F	+20°F	+15°F	+25°F	Freezes at 53.6°F	Freezes at −60°F
Visc., Cp						
70°F		190	4000	10,000	4.5	20.0
300	1.9	2.4	7.4	7.0	.6	1.0
600	.25		1.0	1.0	.35	Not used
700	Not used	Not used	Not used	Not used	.30	
Compatible with materials	Ferrous and nonferrous. Oil-resistant gaskets and packing	Avoid copper and copper alloys. Oil-resistant gaskets and packing	Similar to other mineral oils	Similar to other mineral oils	Carbon steel, cast iron. Aluminum foil packing	Carbon steel, cast iron
Op. Temp./P						
70°F	None required	None required	None required	None required	None required	None required
300	None required	None required	None required	None required	None required	
600	None required	None required	None required	None required	56.3 psig	
700	Not used	Not used	Not used	Not used	120.0 psig	

TABLE 10 Properties of Some Heat Transfer Fluids (continued)

Property	Polyalkylene Glycols (Union Carbide)				Aromatic Base Fluids (Monsanto Chemical Co.)				
	UCON HTF-30	UCON HTF-14	UCON HTF-10	UCON HTF-L20	Therminol 44	Therminol 60	Therminol 66	Therminol 55	Therminol 88
Chemical stability	Stable Water soluble	Stable Water soluble	Stable Water soluble	Stable	Stable	Stable	Stable	Stable	Stable
Oxidative	Up to 150°F in air	Up to 150°F in air	Up to 150°F in air	Up to 150°F in air	Up to 150°F in air	Up to 150°F in air	Up to 150°F in air	Up to 150°F in air	High resistance
Thermal stability	Thermally stable up to 500°F	Thermally stable up to 500°F	Thermally stable up to 500°F	Thermally stable up to 500°F	Thermally stable up to 425°F	Thermally stable up to 600°F	Thermally stable up to 650°F	Thermally stable up to 575°F	Thermally stable up to 800°F
Max. Temp.									
Film	565°F	565°F	470°F	565°F	475°F	635°F	705°F	635°F	850°F
Bulk	500°F	500°F	400°F	500°F	425°F	600°F	650°F	600°F	800°F
Safety									
Fire Res.	Burns	Burns	Burns	Burns	Burns	Burns	Burns	Burns	Burns
Explosive	Negligible	Negligible	Negligible	Negligible	Negligible	Negligible	Negligible	Negligible	Negligible
Density, lb/g									
70°F	8.97	8.63	8.55	8.26	7.75	8.27	8.43	7.40	9.40
300	8.22	7.84	7.76	7.46	6.85	7.60	7.59	6.69	8.42
500	7.50	7.01	7.05	6.76	Not used	6.95	6.75	6.07	7.67
600	Not used	Not used	Not used	Not used	Not used	6.65	6.42	5.76	6.58, 800°F

Spec. heat									
70°F	.44	.44	.44	.44	.46	.383	.365	.460	.467
300	.54	.54	.54	.54	.54	.495	.480	.572	.525
500	.64	.64	.64	.64	Not used	.593	.580	.670	
600	Not used	Not used	Not used	Not used	Not used	.643	.630	.718	.613, 800°F
Therm. cond.									
70°F	0.121	0.115		0.099	0.083	0.076	0.071	.0790	.0712
300	0.099	0.095		0.096	0.071	0.0705	0.067	.0724	.0686
500	Not used	Not used	Not used	0.092	Not used	0.0654	0.064	.0661	.0608
600				Not used	Not used	0.0630	0.062	.0630	.0608, 800°F
Pour point	0°F	−35°F	−45°F	−40°F	−80°F	−90°F	−18°F	−40°F	Melts at 293°F
Visc., Cp									
70°F	320.0	200.0	90.0	200.0	6.20	9.03	142	387	1.57
300	5.4	5.4	3.8	5.4	.44	.87	1.55	1.69	.55
500	1.85	1.85	1.5	1.85	Not used	.38	.45	.58	.23, 800°F
600	Not used	Not used	Not used	Not used	Not used	.29	.34	.42	
Compatible with materials	All common metals inc. copper	All common metals inc. copper	All common metals inc. copper	All common metals inc. copper	All common metals	All common metals	All common metals	All common metals	All common metals
Op. Temp./P									
70°F	None required	None required	None required	None required	None required	None required	None required	None required	None required
300	None required	None required	None required	None required	None required	None required	None required	None required	None required
500	None required	None required	None required	None required	Not used	None required	None required	None required	None required
600	Not used	Not used	Not used	Not used	Not used	None required	None required	None required	None required

Regardless of the pump selected, the flow rate should be checked regularly against the pump's performance when new. To avoid shaft trouble and seal leakage, it is important to avoid pipe support stresses on the pump case. Each pump should be fitted with a control device to switch off the burner in case of pump failure.

If expansion loops are used in the pump section piping, they should be horizontal or vertically downward. Loops should not be vertically upward because this forms a trap that can collect air and vapor, which seriously hamper pump performance.

Carbonaceous deposits can also occur at the pump seal face from leakage at high temperatures. A seal flush system helps prevent excessive seal wear.

Filters

Filters, such as a wire mesh strainer should be installed in a new system. When operating where solids and contaminants might enter the system, it is advisable to install permanently a high temperature filter bypass line that can be isolated with valves for periodic cleaning.

Materials

Most materials normally used in high temperature systems can be used, except for the heater coils. Construction materials are generally selected on the basis of suitability for operation at the specified maximum temperature. Mild steel is widely used. Copper, aluminum, bronze, brass alloys, and the like should be used minimally, mainly because they lose mechanical strength at higher temperatures. Seamless steel tubes selected in accordance with applicable rules of the ASME Boiler Code, Section I, must be used in the heater construction.

Expansion Tank

The expansion tank should be installed at the highest point in the system and connected to the suction side of the pump. The tank should be the main venting point of the system. The double-drop-leg expansion tank provides greater flexibility of operation. Venting of incondensables, water, and so on is often difficult in a heating system. Purging air on start up can also be a problem. The double-drop-leg arrangement insures flooded pump suction and uninterrupted flow on start up and allows venting as needed.

To prevent oxidation, it is preferable when designing the expansion tank to minimize air contact with fluid at temperatures exceeding 100°F. The most effective way to prevent fluid oxidation is to blanket the system with an inert gas such as nitrogen. If this method is not desirable, air contact can be minimized by a cold seal trap arrangement. Low boilers and moisture can collect in the cold seal trap, and the fluid in the trap should be periodically discarded.

Some systems operate with no allowance for oxidation protection of the fluid in the tank. In this case, the temperature in the expansion tank should be kept as low as possible to minimize problems that could result from fluid oxidation.

These guidelines are to assure maximum fluid life, which is particularly important in the case of the higher-cost fluids. When using a nitrogen blanket, moisture should be driven from the fluid before gas pressure is set. Any vent lines should be led outside the building so that vapors do not enter the working area.

The expansion tank should be constructed so that it is one-fourth full when the system is at room temperature and three-fourths full when the system is at operating temperature. It should be fitted with a sight glass at the full range and with a float-operated, minimum-level switch to shut off the heater in the event of accidental fluid loss.

Pipework

The most important factors in the piping layout for fluid systems are (1) proper sizing for the required flow rate and (2) minimizing pressure drop. Because the system undergoes temperature changes, adequate loops to re-lieve expansion and contraction stress are essential. Schedule 40 seamless carbon steel pipe should be used throughout the system. Most organic heat transfer fluids have a tendency to leak through joints and fittings at high temperatures unless these fittings are very tight. The best way to prevent piping leakage is to weld all connections. Where access is necessary, use raised-face flanges with appropriate gasketing and high temperature bolting. Control of piping leaks is especially important, since fluid-soaked insulation poses a more serious hazard than the leaking fluid itself.

When a flanged connection is made, USAS 300-lb weld-neck or equivalent raised-face flanges are recommended. The flanges should be back welded to the pipe and proper gasketing used, such as spiral-wound asbestos and stainless steel gasket. Recommended flange gasketing for high temperature heat transfer fluid systems should be spiral-wound type conforming to API Specification 601. Standard materials for spiral-wound flange gaskets are Type 304 stainless steel and asbestos. Important to nonleak performance with spiral-wound gaskets is the use of raised-face flanges, alloy steel bolting, and even compression of gasket during bolt pull-up.

Insulation

Organic heat transfer fluids exhibit a slow oxidation reaction with the air trapped inside the voids of the insulating material when system temperatures reach about 500°F. Saturated insulation offers a large fuel surface in the face of poor heat dissipation conditions, and this, along with possible catalysis from the insulating material (magnesia, silicate-bonded asbestos, or calcium silicate) can cause a temperature build-up in the mass. This can result in ignition of the fluid when the space between the piping and the saturated insulation is exposed to air. This phenomenon is not fully understood but appears to occur less with closed-cell insulation than with the insulating materials mentioned above. Closed-cell materials thus are preferred, especially where leakage is a possibility despite all precautions.

The problem areas usually are near instrument connections, valve packing glands, flanges, and cracks developing in relatively impermeable insulation cement adjacent to oil-soaked insulation. Any source of leakage should be eliminated promptly. Replace leaky gaskets and oil-soaked insulation, and repack valve stems. Cover insulation where leaks might occur with a hydraulic setting, oil-resistant cement. Install valves, where possible, with stems in a horizontal position so that leaks will drip away from the insulation.

Packing

Various types of packing are used to seal valve stems and small pump shafts on high temperature fluid systems. Flexible metal or solid graphite packings give the best service. Graphite- or Teflon-impregnated, short-fiber asbestos gasketing is also used, but leak-free service is not always achieved at the highest temperatures. Generally, five rings of packing are specified on valve stems to assure a reasonable seal. The use of metal-bellows-stem seals on valves is increasing and, of course, provides absolutely leak-free service.

Valves

Cast steel valves with deep stuffing boxes are satisfactory for organic fluid systems. If possible, globe valves with an outside screw should be used throughout the system since they provide the only truly effective seal against hot fluids. Gate valves do not always provide a tight shutoff.

Controls

Controls for organic fluid systems should be installed both on the heater itself and on the heat-using units. A wide variety of thermal operating controls are available, and any reliable standard equipment is satisfactory.

Install heater controls to regulate the firing mechanism (or waste gas bypass in a heat recovery unit) in direct proportion to the required output. These controls should increase or decrease heat input to maintain the fluid at the operating temperature required by the heat demand of the heat users.

Small units may be operated satisfactorily by "on-off" or "high-low" controllers. Larger units operate better with a modulated control of firing rate.

User controls must be installed to regulate the flow of the heat transfer fluid in proportion to the heat absorption of the user. In multiple-user systems, separate controls must be installed on each user unit to assure proper heat delivery to each. A typical multi-user fluid system, illustrated by Fig. 46, shows a system with a nitrogen charged expansion tank.

FIRE PROTECTION

Selection and sizing of fire protection equipment is an important consideration. It is generally best to consult the client's insurance company for guidance on fire protection equipment. It is also desirable to consult fire safety equipment suppliers.

A commonly used method of fire prevention in the event of tube rupture in fired heaters is piping steam or CO_2 as a snuffer into the heater combustion space. Usually a properly sized line with a remote manual valve for activation is the preferred arrangement. The snuffer system can also be automated by an exhaust stack temperature switch, which energizes a solenoid valve and alarm upon excessive temperature rise, automatically flooding the chamber with a fire extinguishing agent.

In practice, this temperature switch is usually set 100°F over the normal maximum exhaust temperature to prevent nuisance shutdowns. Where steam snuffer systems are used, it is wise to use a steam trap to avoid slugging the combustion chamber with water when the system is activated.

SAFETY CONTROLS

In addition to activating controls, the system must also be fitted with the proper safety devices to meet local code requirements. Safety controls should include the following:

1. High temperature cutoff at the heater outlet to shut off the burner in case of excessive temperature rise. In heat recovery units, the supply of hot gas is diverted from the heater to bypass.
2. A flow sensing switch to activate at a predetermined low flow will shut off the burner or, in heat recovery units, bypass the hot gas to atmosphere. Low flow may be caused by pump malfunction, a leak, or other loss of fluid flow.

    In fuel-fired heaters, burners should be equipped with regular automatic ignition controls and flame failure controls. In wide-range firing, an overfire draft control will save heat losses. Electric power failure and

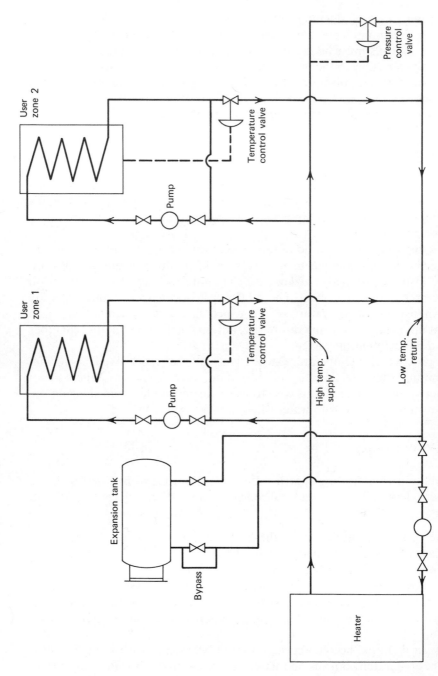

Fig. 46. Organic heat transfer system schematic, with several users at different temperatures.

instrument air failure safety controls are also desirable. In general, a policy of "fail-safe" instrumentation and control in designing is essential, using good quality indicating and recording gauges with accurate reading scales calibrated for the operating range.

Insurance underwriters such as Factory Mutual and Fire Insurance Associates should be consulted for their recommendations.

3. An expansion tank, low level switch must be provided to shut off the heater (or bypass the hot gas) in the event of accidental fluid loss.

HEAT TRANSFER COEFFICIENTS

To determine overall heat transfer coefficients for the heating system, individual coefficients have to be calculated for the heat donor side and the fluid side of each user station. Film coefficients for heat transfer fluids can be calculated using the Seider & Tate-type equation (by Union Carbide Corporation):

$$\frac{hD}{k} = 0.022 \left[\frac{DG}{\mu}\right]_b^{0.8} \left[\frac{Cp\mu}{k}\right]_b^{0.4} \left[\frac{\mu_b}{\mu_w}\right]^{0.16} \tag{35}$$

where $\dfrac{DG}{\mu}$ = Reynolds number (dimensionless)

$\dfrac{Cp\mu}{k}$ = Prandtl number (dimensionless)

$\dfrac{hD}{k}$ = Nusselt number (dimensionless)

$\dfrac{\mu_b}{\mu_w}$ = Viscosity correction factor

h = Fluid film heat transfer coefficient

D = Internal tube diameter, ft

k = Thermal conductivity of fluid, Btu/(h)(sq ft)(F°)(ft)

G = Mass velocity of fluids, lb/(h)(sq ft)

Cp = Specific heat of fluid at bulk temperature

μ = Absolute viscosity of fluid, lb/(h)(ft)

μ_b = Fluid viscosity at bulk temperature, lb/(h)(ft)

μ_w = Fluid viscosity at tube wall, lb/(h)(ft)

b = Bulk fluid temperature (subscript)

Equation 35 may be used for any fluid when its characteristics are known. If the fluid is heated by hot air or by fuel combustion, use Figs. 18 and 19 for gas-side coefficients.

Some producers of heat transfer fluids provide heat transfer data for the fluid-side, which is much easier to use than the equation. Some of these data are illustrated by the following:

Fig. 47 Monsanto Therminol 44

Fig. 48 Monsanto Therminol 55

Fig. 49 Monsanto Therminol 60

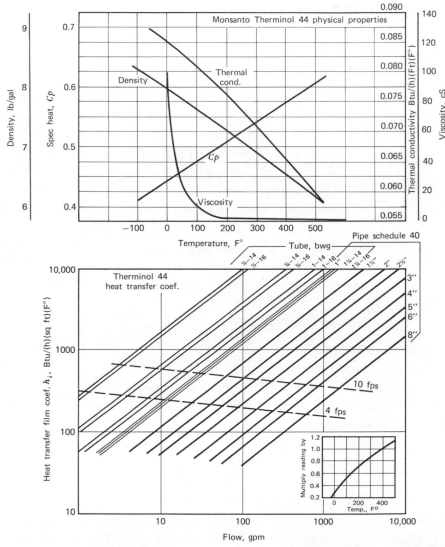

Fig. 47. Heat transfer coefficient, Monsanto Therminol 44. Use range—50 to 425°F; maximum film temperature 475°F.

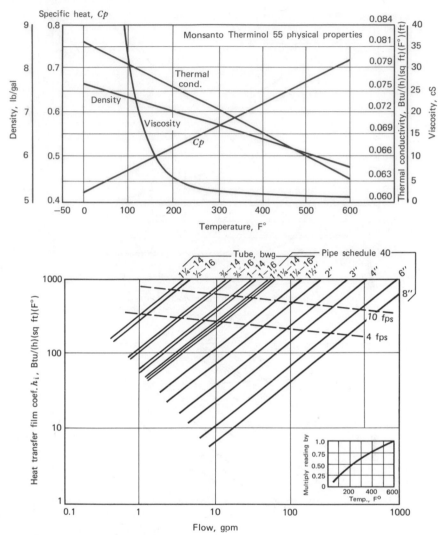

Fig. 48. Heat transfer coefficient, Monsanto Therminol 55. Use range 0°F to 600°F; maximum film temperature 635°F.

Fig. 50 Monsanto Therminol 66
Fig. 51 Union Carbide UCON HTF-14
Fig. 52 Correction factors for Fig. 51
Fig. 53 Dow Chemical Dowtherm "G"

The average fluid velocity used in practice in tubular-type heaters is 4 to 12 fps. It is undesirable to use less than 4 fps because maximum allowable

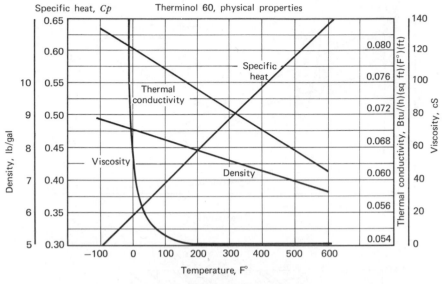

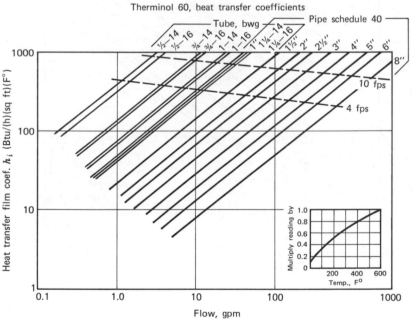

Fig. 49. Heat transfer coefficient, Monsanto Therminol 60. Use range—50°F to 600°F; maximum film temperature 635°F.

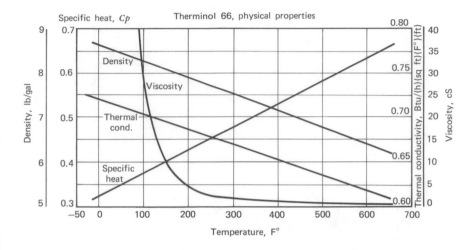

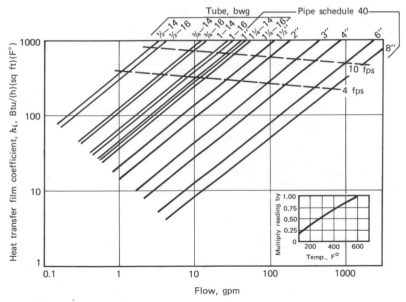

Fig. 50. Heat transfer coefficient, Monsanto Therminol 66. Use range +20°F to 650°F; maximum film temperature 705°F.

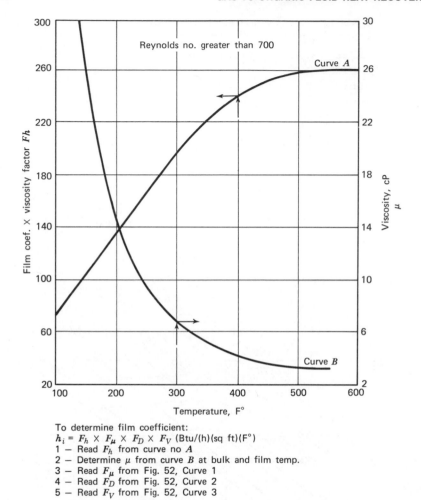

Fig. 51. Heat transfer coefficient, Inside tubes. Union Carbide UCON HTF-14.

fluid film temperatures must not be exceeded since fluid deterioration could result. The upper limit of velocity is actually the maximum allowable pressure drop for the heater, which is usually between 5 and 12 psi.

The determination of maximum fluid film temperature is very important in the successful use of organic heat transfer fluids. Knowing the maximum internal and external coefficients of heat transfer, Equation 2 and Fig. 24 may be used to find the average tube wall temperature at the hotter end. Then, knowing the maximum rate of heat transfer at the hotter end, the tube wall temperature drop Δt_w can be determined. Next, assuming the tube wall temperature gradient is linear, add $\frac{1}{2}(\Delta t_w)$ to t_w, found by Equation 20. The result is inside tube wall temperature and fluid film temperature. This is usually sufficiently accurate for practical purposes. The assumption is that t_w is linear across the tube wall, which it is not, but since tube wall thicknesses are usually not great in tubular equipment (0.085 to 0.125 in. and of high thermal conductivity), the variance from linear may be ignored unless the result is marginal.

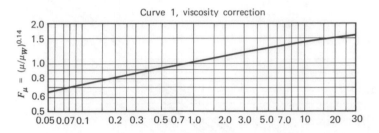

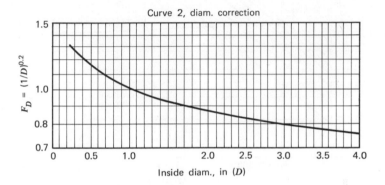

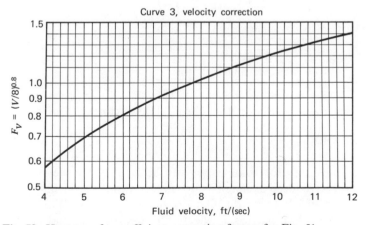

Fig. 52. Heat transfer coefficient correction factors for Fig. 51.

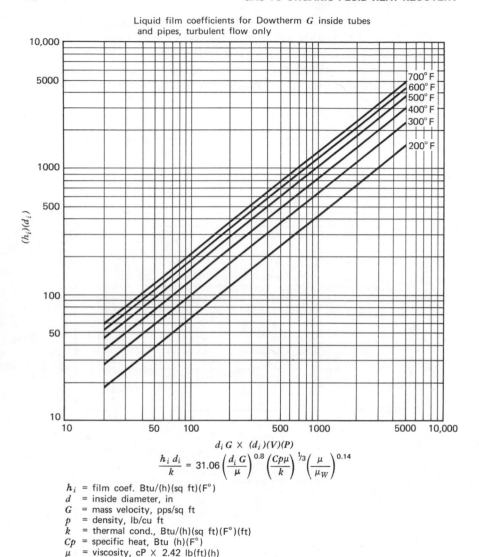

Fig. 53a. Heat transfer coefficient, Dow Chemical Dowtherm "G." Turbulent flow only.

PRESSURE DROP

Fluid-side pressure drop is an important factor in system design. Pumping power must be reduced as much as possible without sacrificing system performance and life. Velocity through the heater is a most important factor because, previously discussed, it affects film temperature and heat transfer coefficient. Velocity through piping can be reduced to less than 4 fps for power conservation; however, larger piping and valves result. These factors must be carefully weighed before making a choice. Velocity through the user equipment is important only in that higher velocities produce higher heat transfer coefficients. Probably a good compromise is 5 to 8 fps through the heater, 4 fps in piping, and then 4 to 6 fps through the user equipment.

Fluid-side pressure drop can be determined from Darcy's equation:

$$h_L = 0.1863 \frac{f L v^2}{d} \qquad (36)$$

where h_L = head loss, ft of liquid
$\quad f$ = friction factor
$\quad v$ = fluid velocity, fps
$\quad d$ = tube i.d., in.
$\quad L$ = length of tube

Friction factor is found in Fig. 54. The Reynolds number can be found by the equation

$$R_e = 6.31 \frac{W}{d\mu} \qquad (37)$$

where W = fluid flow, lb/h
$\quad d$ = tube l.d., in.
$\quad \mu$ = viscosity, Cp

The Reynolds number is then applied to Fig. 54 using the correct tube diameter curve to find the value of f. The viscosity of the fluid must be found in producer's literature on the fluid used and evaluated at the average temperature in the heating coil, or in the case of the piping system, the bulk temperature.

Some fluid producers have developed pressure drop curves for their fluids that are easy to use. The following curves are reproduced here for designer convenience:

Fig. 55 Monsanto Therminol 44
Fig. 56 Monsanto Therminol 55
Fig. 57 Monsanto Therminol 60
Fig. 58 Monsanto Therminol 66
Fig. 59 Dow Chemical Dowtherm "G"

As an example of a problem in an organic heat transfer fluid system, assume that the application is an offshore drilling and production platform and that organic heat transfer fluid is required at a temperature of 550°F for crude oil heating, 450°F for indirect steam generation, and 220°F for heating crew's quarters. The loads are as follows:

Glycol stripping—2,000,000 Btu/h
Crude oil heating—4,000,000 Btu/h

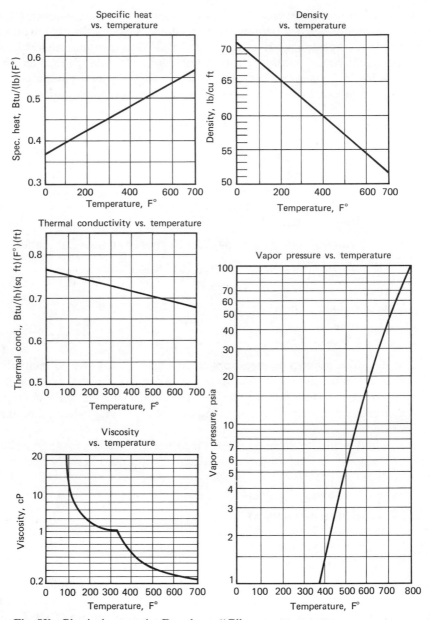

Fig. 53b. Physical properties Dowtherm "G" versus temperature.

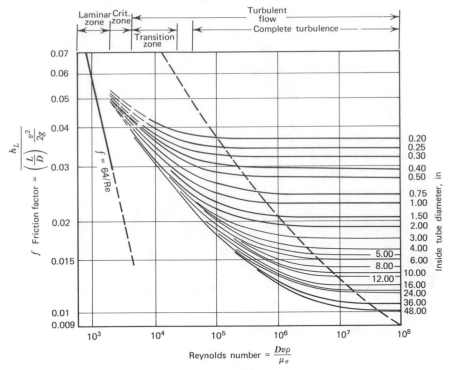

Fig. 54. Friction factor versus Reynolds numbers for liquids.

Steam generation—1,500,000 Btu/h
Crew's quarters—1,500,000 Btu/h
Total load   9,000,000 Btu/h

Figure 46 can be expanded to four users, each operating at a different temperature. Assume further that the platform is in the arctic and requires a fluid that will not freeze at minus 40°F, with an upper limit at least 50°F in excess of 550°F. Referring to the characteristic curves of the several available fluids, consider Fig. 49, which illustrates the heat transfer characteristics of Monsanto Therminol No. 60. This fluid has a working range of minus 50 to 600°F and a maximum film temperature of 635°F. Therefore, Monsanto No. 60 is ideally suited to the application.

Assume that the platform contains several gas turbines, some driving electrical generators, and some driving gas compressors. There are two Solar Centaurs driving gas compressors, which operate at substantially constant load at or near maximum continuous. The two turbines that drive the gas compressors can be used as a waste heat source and can be exhausted into

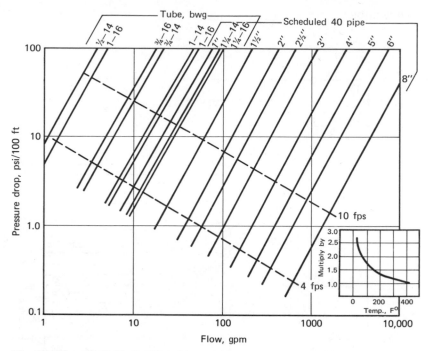

Fig. 55. Therminol 44, pressure drop.

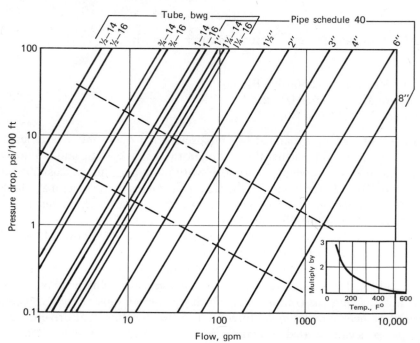

Fig. 56. Therminol 55, pressure drop.

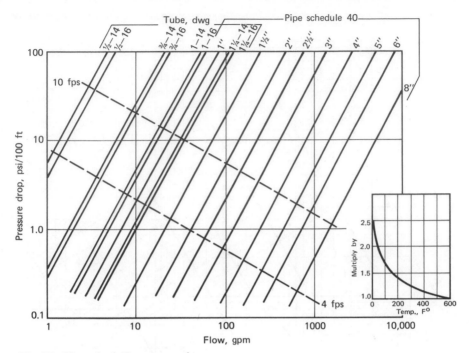

Fig. 57. Therminol 60, pressure drop.

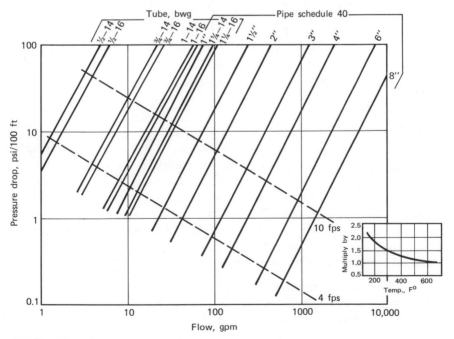

Fig. 58. Therminol 66, pressure drop.

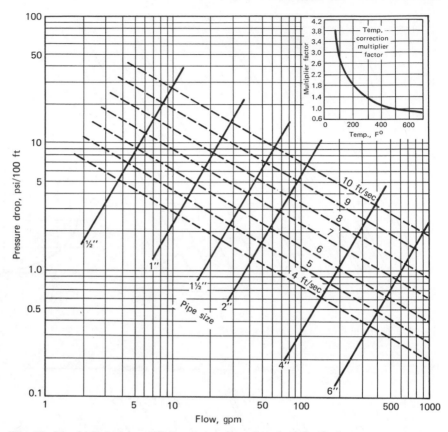

Fig. 59. Liquid Dowtherm "G," pressure drop. In schedule 40 pipe.

one heat recovery fluid heater. The heater can be designed to operate at full thermal load when receiving heat from either of the two turbines. The turbine not used in heat recovery can be exhausted to atmosphere through a bypass valve and duct. Therefore, the heat recovery unit will have two controllable bypass systems, one for each gas turbine, and can be operated from either. The heat recovery unit should be designed to deliver full-rated heat load when the gas turbine is operating at full load, at minus 20°F ambient, sea level. Design specifications are as follows:

Gas turbine power output—5000 shp
Gas turbine exhaust gas flow—44 pps
Gas turbine exhaust temperature—800°F

Heat transfer fluid temperature out—550°F
Heat transfer fluid temperature return—345°F
Average fluid temperature—447.5°F
Average specific heat (Fig. 49)—0.57
Heat load—9,000,000 Btu/h
Required fluid flow rate—77,022 lb/h
Density at 447.5°F—7.1 lb/gal
GPM flow rate—181 gpm

With a fluid temperature into the heater of 345°F, the nearest practical temperature difference between fluid and heater gas exhaust can be taken at 70°F. Therefore, assume the heater stack temperature to be a minimum of 415°F. However, this must be checked against the available ΔT of the gas turbine exhaust when operating under the preceding condition. In determining ΔT available, assume at least a 3% leakage loss through the gas bypass valve when in the fully closed position.

$$\text{Gas } \Delta T = \frac{9,000,000}{(44.0)(.97)(3600)(.257)} = 228°F$$

(The average specific heat of turbine exhaust, from Fig. 2., is .257)

Fluid heater exhaust temperature = $800 - 228 = 572°F$
Lm $\Delta T = 238.6°F$

Fluid velocity should be sufficiently high to prevent excessive local film temperatures, yet with reasonable fluid pressure drop. A minimum velocity of 4 fps is recommended. Tube sizes should be small enough to permit mechanical cleaning should carbonization occur, and not so large as to reduce heat transfer and, therefore, increase unit size and weight. A tube size of $1\frac{1}{4}$-in. o.d. is reasonable for this service. Tube wall thickness must be calculated by Formula 1 in the ASME Boiler Code, Section VIII, for unfired pressure vessels, with P equal to the equivalent pressure for saturated steam at a temperature of 550°F. This pressure is 908 psig. The formula is

$$t = \frac{PR}{SE - 0.6P} \tag{38}$$

where P = design pressure, 908 psig
R = inside radius, $D/2 - t = .625 - t$
S = allowable stress from Table UC3-23, Section VIII, ASME Code for seamless tube specifications. SA179 carbon steel at temperatures to 650°F, 11,700 psi
E = longitudinal joint efficiency—1.0
t = minimum wall—0.047 in.

Use $1\frac{1}{4}$-in. o.d., 16-gauge (.065 minimum wall) tube.

From Fig. 49, $1\frac{1}{4}$-in. o.d., 16 gauge wall tube operating at a velocity of 4 fps requires a flow rate per tube of 11 gpm. A total fluid flow rate of 181 gpm at 11 gpm per tube requires (16.45) use 17 parallel circuits.

The fluid-side coefficient of heat transfer h_i from Fig. 49 is 320. The temperature correction factor for an average fluid temperature of 447.5°F is 0.85. The corrected fluid-side coefficient of heat transfer is then (320)(0.85) = 272 Btu/(h)(sq ft)(F°).

At this point, the gas-side conditions must be determined. It is almost always profitable to use fin tubes on gas turbine exhaust heat recovery applications (see Fig. 37). Consideration must be given to required fin height, heating surface, and gas pressure drop. The latter is important because a power loss of approximately 11 shp is taken for each inch of exhaust pressure drop. The gas pressure drop limitation for most heat recovery applications is usually 6 in. Figure 39 illustrates gas pressure drop through various portions of a typical heat recovery unit. For the sake of expediency, assume that total gas pressure drop through the gas inlet, bypass valve, and gas exhaust parts is 3.00 in. This leaves a pressure drop of 3 in. for the heating surface only. Usually a unit of this type has 5 gas passes. According to Fig. 39, 5 turns (passes) at a pressure drop of 3 in. requires a mass flow rate of 1.80 pps/sq ft. In order to produce this rate, a net free flow (gas passage) area of 44/1.80 = 24.44 sq ft is required. The net gas passage area per parallel circuit = 24.44/17 = 1.438 sq ft.

Next refer to standard fin tube tables such as those published by the Escoa Fintube Corporation and make several trial assumptions for fin configuration on the $1\frac{1}{4}$-in. o.d. tube to produce 1.438 sq ft net free flow area per tube circuit. A circuit length of 75 ft formed into 5 gas passes, of $1\frac{1}{4}$-in. o.d. tube with fins $\frac{3}{4}$ in. high, 7 fins per in., fin thickness of 0.035 in. in carbon steel, produces an average gas free flow area of 1.455 sq ft when fins are at nominal height and spaced 1/32 in. fin tip to fin tip. The external surface per ft of tube is 4.53 sq ft. Total surface for 17 circuits, each 75 ft long, is 5776 sq ft. Next determine whether this arrangement is sufficient for the required heat transfer.

The internal coefficient was 272 Btu/(h)(sq ft)(F°) = h_i (from Fig. 49). The external coefficient is determined as follows:

Mass flow G = 1.80 pps/sq ft. From Fig. 18, for tube diameter D = 1.25,

$$\frac{G^{(.600 + .08 \ ^0 D)}}{D^{.53}} = 1.30$$

Gas temperature in = 800°F

Gas temperature out = 572°F

Average gas temperature = 686°F = 1146°R

From Fig. 19, $T_{av} = 1146$, $T^{1/3} = 10.5$

From Fig. 18, $h_o = (1.30)(10.5) = 13.65$ Btu/(h)(sq ft)(F°)

From Fig. 42, $h_o = 13.65$, $\frac{3}{4}$-in. high fins, and $k \times T = .875 = (25 \times .035)$, $E = 0.67$ fin efficiency.

Ratio of external surface to internal surface $= \dfrac{A_o}{A_i} = \dfrac{4.53}{0.290} = 15.62$

Average tube wall thickness $= 0.0695$ in.

$$R_o = \frac{1}{(h_o)(E)} = \frac{1}{(13.65)(.67)} = 0.10934$$

$$R_i = \frac{A_o/A_i}{h_i} = \frac{15.62}{272} = 0.05743$$

$$R_w = \frac{1}{\dfrac{k/t}{12}} = \frac{1}{4316} = 0.00023$$

$R_{fo} = $ external fouling factor $= 0.00200$ (assumed)

$R_{fi} = $ internal fouling factor $= 0.00100$ (assumed)

$R_o + R_i + R_w + R_{fo} + R_{fi} = R_t = 0.17000$

$$U_o = \frac{1}{R_t} = \frac{1}{0.170} = 5.88 \text{ Btu/(h)(sq ft)(F°)}$$

Thermal load $= 9,000,000$ Btu/h

Lm $\Delta T = 238.6$°F

Required surface $= \dfrac{9,000,000}{(238.6)(5.88)} = 6,415$ sq ft

We found that the proposed tube bank contains only 5776 sq ft. Therefore, another trial must be made. Try the same tube except with fins .025 in. thick, 10 fins per in., $\frac{3}{4}$ in. high, with external surface of 6.02 sq ft/ft. The free flow gas passage area for a 76.5-ft-long circuit in five gas passes is 1.434 sq ft. Seventeen circuits, each 76.5 ft long, will have a total surface of 7829 sq ft. This must now be checked.

The internal coefficient remains as before, 272 Btu/(h)(sq ft)(F°). The external coefficient is

$h_o = 13.65$ Btu/(h)(sq ft)(F°), as before

From Fig. 42, for $h_o = 13.65$, $\frac{3}{4}$-in. high fins, and $k \times T = 0.625 = (25 \times .025)$, $E = 0.61$ fin efficiency

The ratio of external to internal surface $= \dfrac{5.45}{0.290} = 18.79$

$$R_o = \frac{1}{(13.65)(0.61)} = 0.120098$$

$$R_i = \frac{18.79}{272} = 0.069080$$

$$R_w = 0.000230$$

$$R_{fo} = 0.002000$$

$$R_{fi} = 0.001000$$

$$R_t = 0.192408$$

$$U_o = \frac{1}{R_t} = 5.2 \text{ Btu/(h)(sq ft)(F}^\circ\text{)}$$

$$\text{Required surface} = \frac{9,000,000}{(238.6)(5.2)} = 7254 \text{ sq ft}$$

The proposed tube bank contains 7829 sq ft, which is 7.9% in excess of minimum requirements. This is acceptable for practical considerations.

The fluid-side pressure drop must now be found.

Flow per tube—10.65 gpm
Average velocity—4 fps
Tube i.d.—1.11 in.
Average fluid temperature—447.5°F
Circuit length—76.5 ft
Effective length including turns—191.0 ft
From Fig. 57, at flow 10.65 gpm, for $1\frac{1}{4}$-in. 16-gauge tube, $P/100$ ft $= 1.2$ psi
Pressure drop $= (1.2)(1.91) = 2.292$ psi

For a total flow of 181 gpm and a reasonable velocity in the inlet and outlet headers of about 2 fps, use 6-in. Schedule 40 pipe for both inlet and outlet headers. This provides good fluid distribution in the inlet header, leading to substantially equal fluid distribution in all 17 circuits. Fluid piping to and from the heater can be 3-in. Schedule 40, for a velocity of approximately 7.8 fps.

There is an additional pressure drop at the entrance and exit from the tubes. For the inlet, use Fig. 45. For $D/d = 6.065/1.11 = 5.46$ (use the ∞ curve) for $v = 4$, head loss $= 0.12$ ft. For the outlet, use Fig. 44. For $D/d = \infty$ and $v = 4$, head loss $= 0.26$ ft. Total head loss $= 0.38$ ft of liquid $= .140$ psi. Total pressure drop is then 2.43 psi, well below the usual limit of 10 psi for a fluid heater. It is obvious that the fluid-side heat transfer coefficient and hence the overall coefficient is increased by the use of either fewer or smaller tubes, creating a higher velocity to come closer to the arbitrary pressure drop limit of 10 psi. However, smaller tubes do not permit internal mechanical cleaning, and fewer $1\frac{1}{4}$-in. tubes result in insufficient heating surface unless tube circuit length is increased.

Gas-side pressure drop was predetermined in the earlier portion of this example where Fig. 39 was used to find the design gas mass flow rate and number of gas passes. These parameters were maintained throughout this example; therefore, gas pressure drop will be approximately 6 in. wc from the heater inlet flange to outlet flange.

FILM TEMPERATURE

Film temperature evaluation is important in an organic heat transfer system. The maximum allowable fluid film temperature for the fluid of our example, namely 635°F for Therminol 60, must not be exceeded. Fluid film temperature will be evaluated at the point of maximum gas temperature and maximum fluid temperature, which is at the fluid outlet end in a counterflow system. The procedure is as follows:

Gas temperature—800°F
Fluid bulk temperature—550°F
Fluid velocity—4 fps
Tube wall average—0.0685 in.
Thermal conductivity—25

The fluid-side coefficient of heat transfer at 4 fps = 320 Btu/(h)(sq ft)(F°). The temperature correction factor for Therminol 60 at 550°F = 0.95 (see Fig. 49). The corrected value of internal coefficient = (320)(0.95) = 304 Btu/(h)(sq ft)(F°).

The gas-side temperature correction factor from Fig. 19 for temperature 1260°R = 10.80. Gas mass flow is the same. Therefore, h_o = (1.30)(10.80) = 14.0 Btu/(h)(sq ft)(F°). See Fig. 18.

From Fig. 24 and Equation 20, the average tube wall temperature t_w = [(304)(550) + (14.0)(800)]/(304 + 14) = 561°F.

Tube wall temperature drop is expressed by Equation 39.

$$\Delta t_w = \cfrac{q/A}{k_w \cfrac{2\pi l}{\log_e \dfrac{d_o}{d_i}}} \tag{39}$$

The overall value of the heat transfer coefficient U_o maximum is found as follows:

$$R_o = \frac{1}{(14.0)(0.61)} = 0.1170960$$

$$R_i = \frac{18.79}{304} = 0.0618092$$

$$R_w \qquad\qquad = 0.0002300$$

$$R_{f_o} + i \qquad = 0.0030000$$

$$R_t \qquad\qquad = 0.1821350$$

$$U_o = \frac{1}{R_t} = 5.49 \text{ Btu/(h)(sq ft)(F°)}$$

This represents the average value of the coefficient taken over the entire finned surface. The required value is the heat transfer rate q/A referred to the outside of the $1\frac{1}{4}$-in. tube. Gas to fluid ΔT at the hot end is $800 - 550 = 250°F$. The fin-side heat transfer rate is $(5.49)(250) = 1372.5$ Btu/sq ft. The value referred to the bare tube $= (5.45/.327)(1372.5) = 22,875$ Btu/sq ft.

$q/A = 22,875$ Btu/sq ft

$k_w = 25$ Btu/(h)(sq ft)(F°)

l = length of tube/sq ft = 3.06

$\dfrac{d_o}{d_i} = 1.25/1.11 = 1.1261$

$\log_e \dfrac{d_o}{d_i} = 0.1187$

$\Delta t_w = 5.65°F$ tube wall temperature drop

Fluid film temperature $= 561 + 5.65/2 = 563°F$

Therefore, the maximum fluid film temperature is safely below the manufacturer's stated maximum of $635°F$.

The same procedure can be used for any other fluid for which heat transfer and pressure drop curves are included herein. For other fluids, obtain these curves from the manufacturer or use Equation 35, using the physical constants contained in Table 10 or manufacturer's data.

The heater surface assembly can be fabricated in either a rectangular or circular configuration. In either case, the effective circuit length is 76 ft 6in. in five gas passes. For the rectangular configuration, the circuit length is made up of five straight lengths, each $185\frac{5}{8}$-in., joined by 180° return bends. Seventeen such elements are required.

For the circular configuration, the circuit length is wound on a horizontal circular turntable to form a spiral coil of five turns. Seventeen such coils are required, stacked one on top of the other, with ends connected into vertical headers, one in the inside of the assembly and one at the outside. Refer to Fig. 36 for a typical completed assembly. The top of the coil bundle is baffled to the central hole. The bottom is baffled clear across the coil bundle. Gas flow is radial from the outside of the coil assembly to the inside, and then out through the stack.

With either heating surface form, fluid temperature is controlled by the turbine exhaust bypass. The bypass valve may be two oppositely operating butterfly valves or one "clam shell"-type, swinging gate valve. The clam shell type gives better control than butterfly valves and can be counterbalanced, thus requiring very little power to operate.

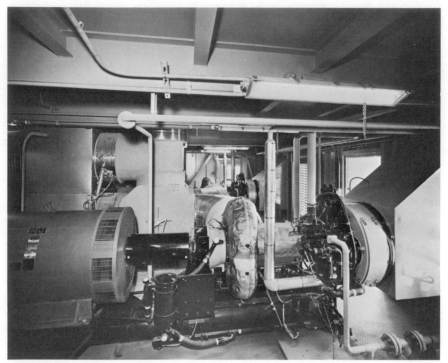

Fig. 60. Gas turbine heat recovery heat transfer fluid heater on offshore platform. Courtesy Hutchison-Hayes, Inc.

Figure 60 is a photo of the actual installation of a fluid heater and two Solar Saturn gas turbines on an offshore platform. The two vertical assemblies in the foreground are the twin bypass assemblies of the clam shell valve. This unit was designed to operate with either of the two turbines or both turbines operated with one flowing through the heater and the other through bypass.

ORGANIC FLUIDS IN POWER CYCLES

Organic heat transfer fluids having temperature entropy characteristics that make them suitable for power cycles have recently been developed. These fluids are generally fluorocarbons. Bottoming power cycles with the gas turbine as the prime power source look especially attractive. producing overall cycle efficiencies of up to 41%. The fluid is heated and vaporized in the gas turbine heat recovery boiler to a pressure of 700 psig and a temperature of 600°F. It is then used in a high speed turbine either directly coupled to the gas turbine generator or driving a separate generator through a reduction gear. Obviously, such a system must be sealed so that the fluid can be conserved.

A more detailed discussion of the organic fluid bottoming cycle is contained in Chapter 5.

HEAT RECOVERY BY STEAM GENERATION

The recovery of heat energy by the generation of either high pressure or low pressure steam is the most common means of fuel and energy conservation. This mode of recovery may take several forms, depending upon the desired result.

LOW PRESSURE

Steam generation at low pressure (10 to 12 psig) is used for space heating and/or absorption air conditioning. The heat source may be the exhaust of gas turbines or the effluent gas from an incinerator. Many total energy systems in operation also generate low pressure steam by heat recovery from a reciprocating engine cooling jacket. In some cases, heat in the engine exhaust is also recovered by a dual heat unit. However, because of the temperature limitation on the engine cooling system, steam generation is limited to 15 psig (250°F).

Heat recovery for low pressure steam generation from the exhaust of a gas turbine is accomplished by either a natural circulation or a forced recirculation boiler sized to match the gas turbine gas flow. Figure 61 illustrates a typical forced recirculation boiler designed to accept flow from two gas turbines. It contains two separate gas bypass control systems, one for each gas turbine. Both gas bypass valves can be modulated simultaneously by a steam pressure controller, or either can be modulated with the other in full bypass, when only one turbine is in operation. The boiler illustrated can generate steam at 6000 lb/h at 12 psig from the exhaust from two 225-kW turbines. This type of forced recirculation boiler contains a heating surface assembly illustrated by Fig. 36.

Other types of heat recovery boilers for this application include units of straight-tube banks attached to fixed or floating headers and units of serpentine (return-bend) elements. The circular coil type of Fig. 36 and the horizontal serpentine elements require forced recirculation. Vertical tube units may operate in either forced or natural circulation modes. Gas turbine exhaust heat recovery boilers of moderate size (to 10,000 lb/h) operating at low pressure seldom have supplementary firing. Stack temperature is usually 70 to 100°F above steam temperature. Typical conditions are as follows:

Turbine exhaust temperature—980°F
Steam pressure—12 psig (240°F)
Boiler stack temperature—320°F

Bypass valves are almost never gas tight. At least 3%, and at times 5%, leakage loss is usually allowed (i.e., when the bypass is fully closed, gas leakage up the bypass stack will be between 3 and 5%). Recovery effectiveness under these conditions is 71%. Heat recovery can be increased by operating at reduced stack temperature, say 40°F, above steam temperature. Heat recovery effectiveness will be 75.4%; however, heating surface will increase exponentially. The effect of stack approach temperature difference on Lm ΔT surface is illustrated by Fig. 62. Obviously, increased boiler cost must be weighed against equivalent fuel cost over time.

Fig. 61. Gas turbine heat recovery boiler. Courtesy Consec., Inc.

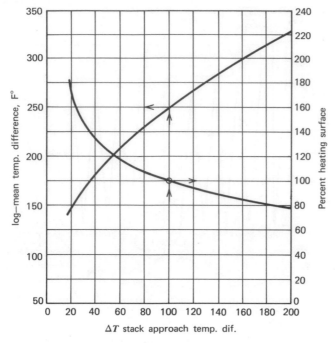

Based on steam temp. 353° F (125 psig)
entering gas temp., 850° F

Fig. 62. Effect of stack approach ΔT on Lm ΔT and heating surface.

Forced Recirculation

Figure 63 is a typical schematic of a forced recirculation steam generation system used on both low and high pressure systems. Pressure rating is limited by the availability and cost of recirculating pumps capable of continuous operation at elevated temperatures. Since forced recirculation may be used at both high and low pressures, a more detailed discussion follows in a later paragraph.

Natural circulation

Larger low pressure heat recovery applications usually employ the natural circulation system, mostly of the two-drum variety. A schematic of such a system is illustrated by Fig. 64. It is essentially a closed loop, with one leg of the loop receiving heat and the other serving as a downcomer that does not

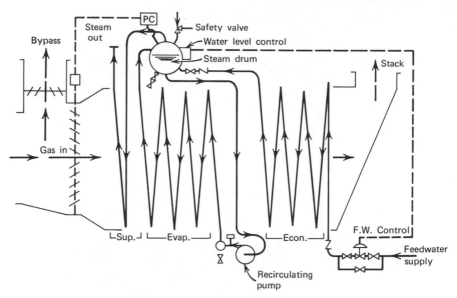

Fig. 63. Schematic, forced recirculation boiler system.

receive heat. The difference in water and steam specific volume in the heated portion and the water only portion in the downcomer creates a natural up-flow of water and steam from the lower to the upper drum. Steam is released in the upper drum, passing through a scrubbing section for moisture removal.

The heat source for a typical low pressure boiler for steam rates in excess of 10,000 lb/h may be either a gas turbine or an incinerator. The incinerator may be (1) a solid waste municipal type or institutional incinerator, (2) a sewage sludge incinerator, or (3) a gaseous or liquid waste incinerator. Boilers for such applications are almost always of the two-drum natural circulation type. Data on an actual sewage sludge incineration low pressure steam generating case follow:

Total dry gas flow—74,682 lb/h
Water vapor flow—18,671 lb/h
Total gas flow—93,353 lb/h
Gas temperature—1200°F
Feedwater temperature—212°F
Steam pressure (operating)—15 psig
Bypass leakage loss allowance—5%
Boiler stack temperature—423°F

Steam rate—20,000 lb/h
Total boiler surface—6775 sq ft
Heat recovery—19,680,000 Btu/h
Average overall rate of heat transfer—2905 Btu/sq ft
Lm ΔT—456.7°F
Average overall U_o—6.36 Btu/(h)(sq ft)(F°)
Heat recovery effectiveness—69.4%

The boiler described above is a two-drum natural circulation boiler, of vertical arrangement with gas flow into the top of one side and out near the bottom of the same side, making one 180° turn inside around an inclined baffle. The tubes are of 2-in. o.d., 0.105-in. wall with external fins $\frac{3}{4}$ in. high, 6 fins per $\frac{3}{4}$ in., 0.050 in. thick. The fins are the segmented type illustrated by Fig. 37. Some designers prefer this type because of increased turbulence and a more open pattern that facilitates cleaning by soot blowing. Another fin

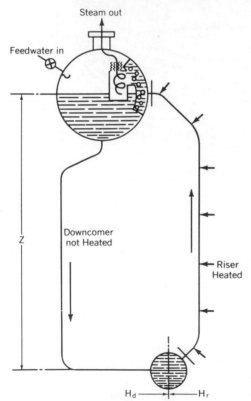

Fig. 64. Schematic, natural circulation boiler. Courtesy Babcock & Wilcox.

form is the solid fin illustrated by Fig. 65. The solid fin is used principally on larger tubes and pipe. Heat recovery effectiveness could increased to 78.6% if the boiler had been designed for a stack temperature of 320°F, which would have been possible with operation at 15 psig. Lm ΔT would have been reduced to 337.8°F, total heating surface increased to approximately 10,898 sq ft, and steam production increased to 23,794 lb/h. An increase of 61% in heating surface results in only a 19% increase in steam production.

Fig. 65. Solid fin tube. Courtesy Escoa Fintube Corporation.

Reciprocating Engine Heat Recovery

The heating surface in a reciprocating engine jacket water heat recovery is the actual surface of the engine cooling jackets. Most systems operate in a natural circulation mode. The steam separator vessel is mounted over the engine to create a substantial head of water on the inlet side of the engine. The outlet end of the jacket is connected to the steam separator inlet nozzle. The differential head created by the all-water downcomer and the steam-water riser induces water circulation through the engine jackets. A portion of the circulated water evaporates, and the resulting steam is released from the steam separator steam nozzle. Unevaporated water is recirculated, and evaporated water is made up in the steam separator by means of the addition of water by the level and feed control system.

Sometimes engine heat recovery systems have an auxiliary radiator for control in cases where all the steam generated by the engine may not be required. With this arrangement, a water-circulating pump is needed to provide the required head to divert some or all of the cooling water to the radiator. When exhaust heat from the engine is also recovered, a heat exchanger is added to an enlarged separator vessel for heat transfer from exhaust gas to water at boiling temperature.

When selecting a heat recovery unit for an engine, the following information is needed:

Engine model no.
Engine speed
Full-load horsepower
Engine jacket heat rejection rate
Engine jacket water rate required
Engine jacket full-load temperature
Exhaust gas flow rate, pps
Exhaust gas temperature
Maximum permissible back pressure

Heat recovery for large reciprocating engines is sometimes similar to a fire tube boiler with the engine exhaust passing through the fire tubes and the engine cooling water taken from the boiler water drum. Often two or more engines may be connected to one boiler. The boiler must be above the engine to provide a head for recirculation. Hot water to the engine is taken from the bottom of the boiler drum, and the water outlet from the engine is taken back to the steam space of the boiler drum. This arrangement provides for natural circulation, often known as ebullient cooling. It is, of course, important to assure that the water level in the drum is always at the normal point, and that the feedwater controls always operate properly; otherwise engine damage could result.

In the fire tube engine heat recovery system, it is a simple matter to add the normal firing system usually associated with a fire tube boiler. This arrangement provides supplementary firing that can be used when the engine load is low but heat demand is high. It also makes possible the total firing of the boiler with the engines completely off, as for engine maintenance conditions.

The detailed design of a fire tube heat recovery system will not be dealt with here because these units are commercially available in a wide range of sizes to suit almost any engine size normally used in total energy systems. However, care must be taken to match boiler characteristics with heat rejection requirements of the engine when operating at overload conditions.

Some details regarding the design of reciprocating engine heat recovery must be observed that are peculiar to that application, including the following:

1. A large volume of water and an ample steam liberation area are necessary to avoid the return of superheated water to the engine jacket, thus avoiding the possibility of steam bubbles forming in the engine jacket, which results in hot spots and possible engine deterioration.
2. There should be a good sedimentation chamber away from the engine to remove contaminants either brought in by the water or by chemicals added to the water for removing corrosive gases and dissolved solids.
3. Gas-to-water heating surfaces must be accessible for easy inspection and repair.
4. Engine exhaust turnaround chambers within the boiler should be designed for minimum pressure drop and to act as pulsation dampeners for exhaust noise suppression.
5. The heat recovery boiler must be furnished with a high level cutoff, high level alarm, low level pump start, low water alarm, separate mechanical water feeder for emergency service, low water cutoff, vent valve, ASME safety valve, pressure gauge, blow down valve, and the required interconnecting piping.
6. The boiler must be properly insulated, covered with a metal jacket. Turnaround chambers are internally insulated for pulsation dampening and noise suppression.
7. Gas entrance nozzles must be constructed of stainless steel to resist high temperature scaling.

Steam pressure available from a reciprocating engine heat recovery system is limited to 15 psig (250°F) because the engine jacket conditions are limited to 250°F operation. Total heat rejection from an engine is about one-third that of a comparable gas turbine. This is, of course, due to its higher operating efficiency. For example, the heat recovery from a 1100-hp gas turbine is about 6,200,000 Btu/h, compared to 2,100,000 Btu/h from a comparable engine with both boilers operating at a stack temperature of 350°F.

The engine-based total energy system enjoys widespread use in the United States accounting for at least 95% of all installations, in spite of the greater reliability of the gas turbine. However, lower capital investment and reduced fuel requirements make it attractive. The gas turbine, with its lower maintenance cost and greater heat recovery capability, can be very attractive in cases where all or most of the recovered heat can be used all the time.

Waste heat recovery boilers used on marine engines of up to 5000 bhp recover heat from the engine exhaust only. Therefore, operating pressure ranges from 10 to 200 psig. One type of marine heat recovery boiler is made up of multiples of flat, spiral coils stacked one on top of the other, with gas flow axially through the coil bank. Control valves at the coil inlets cut coils in and out of the circuit, thereby varying heating surface with load changes. With this type of control system, the entire heating surface must be capable of running dry indefinitely at maximum engine exhaust temperature.

An important feature of spirally coiled tubes is their flexibility, whereby all stresses due to unequal heating are completely relieved. In addition, this type of construction permits engine operation with no water passing through the boiler for steam generation.

An existing auxiliary boiler may be used as a receiver, thereby eliminating the need for a separate drum. This can be done whether the drum is above or below the waste heat boiler, since the circulating pump delivers the steam and water mixture to the receiver, where separation takes place. If there is no suitable auxiliary boiler for connection to the waste heat boiler, the coil assembly can have its own drum with recirculating pump.

The boiler can be constructed so that the inlet and outlet headers are outside the casing, with inspection and clean out plugs opposite each coil terminal to facilitate chemical cleaning.

HIGH PRESSURE

High pressure applications are divided into the following categories: (1) process steam, (2) sewage heat treatment, (3) gas turbine–steam turbine cycles, and (4) municipal waste incineration.

Process Steam

Steam generation at pressures up to 125 psig is used with product steam for process work and in hospital energy systems using higher pressure steam for laundry and sterilizing applications. In such cases, the heat source is exhaust of either gas turbines or various types of incinerators. Boilers for such applications can be either forced recirculation or two-drum natural circulation, as described above. Effluent gas from a solid waste incinerator is usually not clean; consequently, any boiler operating in conjunction with a solid waste incinerator must not contain extended surface or finned tubes. All tubes must be plain, generously spaced, with pneumatic or preferably steam-operated soot blowers arranged to cover all surfaces effectively. The effluent gas must be analyzed for possible corrosives, which will affect the type of material used for the tubes. Waste heat boilers connected to hospital incinerators are particularly vulnerable to attack because of the high organics and plastics content of the refuse.

Sewage Heat Treatment

Saturated steam generated at pressures of from 250 to 350 psig is used in heat treatment of sewage sludge in plants requiring high temperature treatment of sludge. The untreated sludge is pumped through one side of a regenerative heat exchanger to a reactor vessel, where it is held for a predetermined time. Treated sludge flows from the reactor vessel through the hot side of the heat exchanger, thereby heating the incoming sludge and requiring the addition of only the final 5 to 10°F of temperature by direct steam injection to the reactor. Required steam pressure is from 250 to 350 psig, depending on the requirements of the system.

In some sewage heat treatment systems, compressed air is introduced along with the cold sludge, producing an exothermic reaction between the oxygen content of the air and the hydrocarbon content of the sludge. This process produces added heat that reduces the steam requirement to the reactor. Steam required for a high pressure sewage sludge treatment plant is generated by direct-fired boilers backed up by heat recovery boilers deriving their heat input from the sewage sludge incinerators. The treated sludge from the regenerative heat exchanger is dewatered by centrifuges or vacuum filters and conveyed to multiple-hearth furnaces for burning. The furnaces may consist of 7 to 11 hearths. Sludge fed into the top hearth proceeds downward to each succeeding hearth until it is discharged at the bottom as ash. Sometimes the sludge is fed into the second hearth from the top, and the top hearth is used as an afterburner to consume all residual hydrocarbons. For best results, furnace outlet gas temperature must not be less than 1100°F and preferably between 1200 and 1400°F. At these temperatures, it is quite safe to assume that hydrocarbon combustion is complete and that there will be no unburned hydrocarbons carried over with the effluent gas. There will be solids carryover in the form of fly ash, which in well-operated furnaces is a dry residue, ranging from talcum powder consistency to granular consistency similar to table salt. This material can amount to as much as four or five pounds per 1000 cu ft of gas. Obviously, the heat recovery boiler must be designed to handle the solids as well as the hot gas. The heat recovery boilers used in conjunction with a sewage sludge incinerator are usually of the two-drum natural circulation type. No horizontal baffles should be used since they accumulate fly ash, which tends to pile up on them. All baffles must be either vertical or steeply inclined. Boiler tubes are of 2-in. o.d. and are bare for the first two rows, but they may be finned for the remainder of the boiler. If fins are used, it is recommended that they not be over $\frac{5}{8}$ in. high, not closer than 6 fins per in., with a thickness of 0.050 in. Under some circumstances, $\frac{3}{4}$-in.-high fins are used, but this is the absolute maximum. Segmented fins are preferred over solid fins because they open more free lanes for soot blowing.

The placement of rotating and fixed soot blowers is important. Usually at least two rotating soot blowers are required to do the job effectively. In addition, a fixed soot blower should be located at the upper edge of inclined baffles to keep them free of ash and over one edge of the ash hopper to prevent bridging and facilitate the flow of ash down the hopper and out to the ash conveyor.

Effective placement of ash hoppers is important, because ash accumulates quickly if it is not periodically, or preferably continuously, removed. Gas flow into the boiler at the top and out at the bottom, if properly arranged, eliminates the need for ash hoppers and ash conveyors, because the ash will be conveyed through with the gas directly into the scrubber system.

In most cases, the steam generated by heat recovery from the incineration of sewage sludge is sufficient to carry the heat treatment load, with steam to spare for other uses such as building heating. The fired boilers then come on only when heat recovery drops below that required to carry the load. Studies are presently under way to investigate the use of surplus steam by heat recovery from sewage sludge incineration for power generation.

Figure 66 illustrates a typical heat recovery boiler used with a sewage sludge incinerator.

Gas Turbine–Steam Turbine Cycles

Combination gas turbine and steam turbine power cycles are of continuing interest, particularly in these times of fuel shortages and high fuel costs. Combination cycles are being used by electric utilities in sizes of up to 300 MW. Studies are also under way on smaller plants of up to ten MW.

Overall cycle efficiencies for such plants range from 37 to 41%, which makes them quite attractive from a fuel conservation viewpoint. The lead time required for construction and capital cost is less than that required for a conventional steam power plant of comparable capacity.

A combination gas turbine–steam turbine plant consists basically of a gas turbine generator, a heat recovery boiler to recover heat from the gas turbine exhaust, a steam turbine generator driven by the steam generated by the heat recovery boiler, and a steam condenser system.

Unfired heat recovery combined cycle. Heat recovery boilers for straight unfired heat recovery systems usually operate at pressures up to 450 psig and steam temperatures to 650°F. Supplemental-fired and exhaust-fired heat recovery boilers for combined cycles operate at up to 1450 psig at 900°F. Both forced recirculation and natural circulation boilers are used, although larger plants have natural circulation boilers with economizers and superheaters. Figure 67 illustrates a gas turbine exhaust heat recovery boiler that includes a bypass stack and control valves, a grid-type supplementary gas burner, a superheater section, an evaporator section and an economizer, and an exhaust stack.

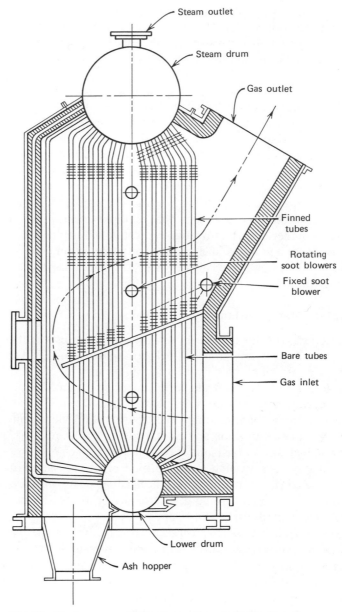

Fig. 66. Heat recovery boiler for sewage sludge incinerator.

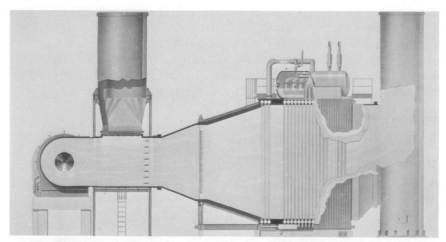

Fig. 67. Heat recovery boiler with bypass, burner, superheater, and economizer. Courtesy Henry Vogt Machine Co.

High pressure heat recovery boilers sometimes have a low pressure section after the high pressure economizer. The low pressure section is actually an additional boiler section because it has its own steam drum, mud drum (or equal) and downcomers. Low pressure steam from this section is used for deaeration and feedwater heating and has a distinct advantage in overall economy. Such a boiler is illustrated by Fig. 68. A typical large gas turbine

Fig. 68. Dual pressure heat recovery boiler. Courtesy Struthers Wells Corporation.

heat recovery boiler is illustrated by Fig. 69. The thermal efficiency of a simple cycle gas turbine is about 25%. Almost all the rejected heat, or 75%, is in the exhaust gas, which is also rich in oxygen. The gas turbine therefore lends itself to the recovery of heat from its exhaust, with or without the combustion of supplementary fuel. The basic combined cycles for electric utility application are (1) the unfired heat recovery cycle, (2) the supplemental-

Fig. 69. Typical gas turbine heat recovery boiler installation, with by-pass and supplementary burner; 150,000 lb/hr at 150 psig, and 30,000 lb/hr at 15 psig. Courtesy Struthers Wells Corporation, Warren, Pa.

fired heat recovery cycle, and (3) the exhaust-fired cycle. These systems all employ gas turbines and condensing steam turbine generators. The basic differences are in the boilers and the proportion of the plant output produced by the steam system. The steam system produces 30 to 35% of plant capability in the unfired heat recovery cycle, 35 to 60% in the supplemental-fired cycle, and 60 to 80% in the exhaust-fired cycle.

The unfired combined cycle is the simplest of the combined-cycle systems, using only the energy in the gas turbine exhaust for steam generation. The temperature entropy diagram of Fig. 70 illustrates the cycle energy utilization

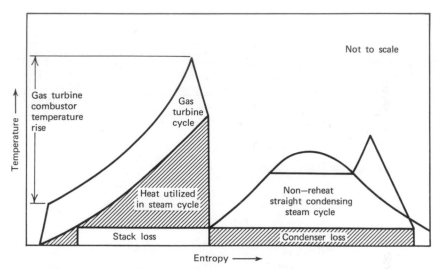

Fig. 70. Temperature entropy diagram for unfired combined cycle.

and heat rejection. A schematic cycle diagram of a typical unfired heat recovery combined-cycle plant is illustrated by Fig. 71.

The steam turbine applied in the unfired combined cycle is usually a straight condensing unit with no feedwater heating. The most efficient cycles use energy from the gas turbine exhaust for performing all feedwater heating duty to achieve minimum stack temperature and maximum efficiency. The steam conditions are modest and usually do not exceed 850 psig, 900°F total temperature.

The feedwater heating system may include a deaerator using steam or hot water heated by heat transfer surface in the low temperature section of the boiler. This is the most common system; however, systems have been installed

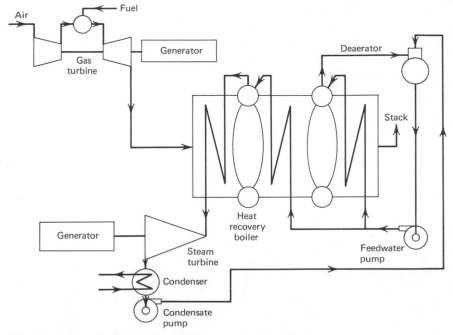

Fig. 71. Schematic, unfired combined heat recovery cycle.

using deaerating condensers, with the deaerating feedwater heater eliminated. Satisfactory operation is achieved with the deaerating condenser cycle because of the modest steam conditions. Deaerating condenser cycles use an enlarged economizer section for all feedwater heating. Low stack temperatures can be achieved by this method. Heat recovery boilers applied to combined cycles usually consist of a superheater, an evaporator section, an economizer, and in some cases, a low pressure boiler section for feedwater deaerating. A low pressure boiler is sometimes included for generating steam that is admitted to the low pressure stages of the steam turbine. This arrangement improves cycle efficiency, particularly when applied with gas turbines having low exhaust temperatures (less than 950°F). Heat recovery boilers having three or four separate heat exchange sections and operating at comparatively low gas temperatures must be designed for low approach temperatures, requiring a large amount of effective heating surface. Extended surface tubes are employed to attain economically the required low approach temperatures.

The characteristic of the unfired combined cycle that enables it to attain low heat rates is the low heat rejection to the condenser, since the steam system output does not exceed 30 to 35% of total plant capability. Net heat rates are in the range of 8000 to 8500 Btu/kWh based on the high heating value of the fuel. Much depends on gas and steam turbine efficiencies and the percentage of the total heat in the gas turbine exhaust recovered as steam.

A low-part-load heat rate can be achieved with gas turbines in which air flow can be modulated in proportion to power output. Air flow modulation is achieved by the use of multiple-shaft gas turbines or modulation of compressor inlet guide vanes on the single-shaft gas turbines. Best results are attained by the use of several gas turbines and one steam turbine. The lowest heat rates occur when the gas turbines are fully loaded; therefore, the most efficient operation is achieved by loading in increments of gas turbine capacity similar to the practice of loading large steam turbines at the best valve points.

Firing all the fuel into the gas turbine eases the operation and adapts it to automated and semi-attended operation, because boiler firing and safety controls are eliminated. The steam turbine should be equipped with an initial pressure governor that maintains steam pressure at the rated level throughout the load range. Bypass stacks and dampers for conducting the gas turbine exhaust gases to atmosphere are commonly used for control of the heat recovery boiler and for start up. Sometimes the system is equipped with a steam dump to the condenser to facilitate steam turbine start up.

One desirable feature of the unfired heat recovery combined cycle, using the single shaft gas turbine with modulated compressor inlet guide vanes, is maintenance of constant steam temperature to the steam turbine over the upper 30% of the load range. This feature eliminates a criticism that has been commonly directed toward the unfired combined cycle, namely, poor steam temperature control with load variation.

Installation costs of unfired heat recovery combined cycles are usually low because a large part of the plant capability is produced by the gas turbines. Figure 72 illustrates the range of installed costs for unfired combined cycles. The cost estimate on which this illustration is based includes evaporative cooling towers, an oil fuel system, and auxiliaries for a continuous service plant. Land and switchyard are not included in the estimate. The wide range of costs results from individual preferences, variations in labor costs, differences in steam cycle design, geographic and climatic conditions, and variations in the costs of equipment from different manufacturers.

Supplemental-fired heat recovery combined cycle. The supplemental-fired heat recovery combined cycle, the most popular among utility users, comprises a gas turbine, a supplemental-fired heat recovery boiler, a non-reheat steam turbine, and a condensing system. Supplemental firing of the boiler

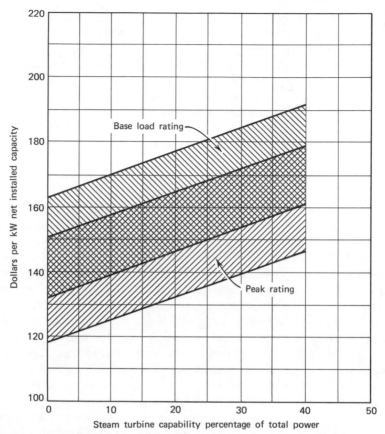

Fig. 72. Range of installation costs for unfired heat recovery combined cycles. Costs projected to 1975 estimates.

increases the output of the steam system over that of the unfired boiler system, and the range of steam system output is 35 to 60% of total plant output. The temperature entropy diagram is illustrated by Fig. 73. A typical cycle schematic is illustrated by Fig. 74.

The supplemental-fired heat recovery boiler is the same as the unfired boiler except for the addition of supplementary burners. Figure 68 illustrates such a boiler. The burner assembly is located upstream of the superheater. The furnace is not heavily fired, so that refractory lining is sufficient. The range of gas temperatures leaving the furnace section and entering the super-heater is usually 1100 to 1400°F. The boilers are equipped with finned tubes

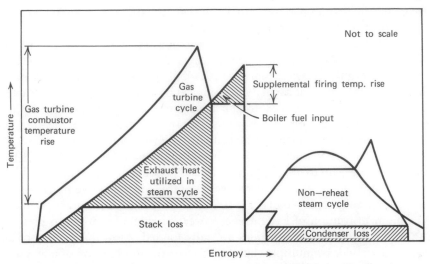

Fig. 73. Temperature entropy diagram for supplemental-fired combined cycle.

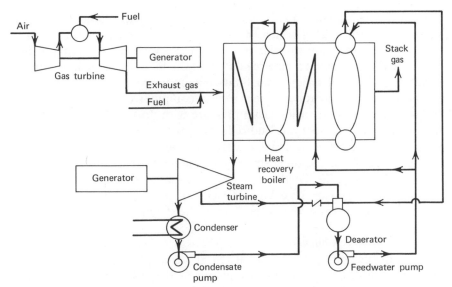

Fig. 74. Schematic, supplementary fired heat recovery cycle.

except for the first four passes, usually the superheater, which are bare tubes, when operating at the higher gas temperatures.

Exhaust gases are used to the greatest possible extent for feedwater heating for maximum efficiency. Supplemental heat is sometimes required from steam extracted from the steam turbine at high loads on the system in which more than about 45% of plant capacity is produced by the steam system. Extracted steam is usually introduced to the deaerating feedwater heater, supplementing the steam from a low pressure boiler for feedwater heating, as illustrated by Fig. 74. Such a cycle is highly efficient, with low heat rejection to the condenser. Net heat rates are in the range of 7800 to 8300 Btu/kWh for plants where up to 45% of total output is produced by the steam turbine.

The supplemental-fired heat recovery combined-cycle system can be designed for excellent part-load heat rate. Good part-load performance requires maximum utilization of gas turbine power, with minimum or no firing of the boiler unless the gas turbines are fully loaded. Also, the control of air flow to the boiler by multiple-shaft turbines or the modulation of inlet guide vanes on single-shaft gas turbine compressors is required for maximum part-load efficiency.

One major disadvantage of the supplemental-fired system is the sharp variation in steam temperature with load, because air flow is not varied with steam flow when the boiler is fired, and the superheater is the element nearest the burners. Some systems have been installed with the superheater upstream of the burner, alleviating this situation. Such an arrangement is particularly applicable to cycles using low temperature steam, approximately 75 to 100°F below gas turbine exhaust temperature. A further disadvantage is the complexity of another firing system with its controls and safety equipment.

Steam pressure is controlled in this system by a wide-range initial pressure governor on the steam turbine regulated on the high end by a conventional combustion control system that limits maximum pressure by modulating the firing rate. Some systems now have a computer control on the steam turbine governor that varies steam pressure with flow. This type of control minimizes part-load fuel rate and reduces moisture in the steam in the low pressure turbine stages during low-load operation.

The range of installation costs for supplemental-fired combined cycles is illustrated by Fig. 75. This type of system usually costs slightly more per kilowatt of base load or peaking capacity than the unfired heat recovery combined cycle because the proportion of the steam capability in the plant is greater.

Exhaust-fired combined cycle. The exhaust-fired combined cycle consists of a gas turbine and a conventional reheat steam plant with the preheater and forced draft fan for the boiler replaced by the gas turbine. This type of system has been the subject of a great deal of study, and several such systems

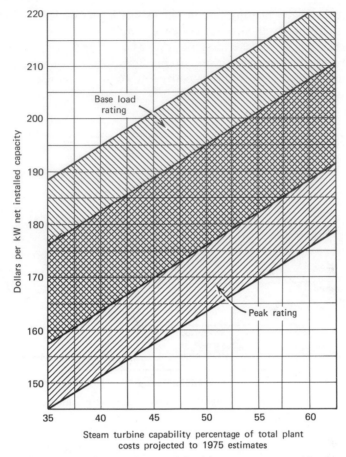

Fig. 75. Installation costs for supplemental-fired heat recovery combined cycle. Costs projected to 1975 estimates.

have been installed. The heat rate of a conventional steam plant can be improved by 6 to 9% using currently available high temperature gas turbines.

Since all stack gas cooling must be performed by feedwater heating, cycles designed for maximum efficiency have complex feedwater heater arrangements. Feedwater heating systems for exhaust-fired cycles have been considered extensively, and two basic types have evolved. One is the parallel feedwater heating system, in which a secondary economizer is installed parallel to the feedwater heaters. The secondary economizer maximizes the feedwater heating duty of the stack gases for maximum efficiency. A less expensive feedwater heating system, the "short heater cycle," is slightly less efficient than the parallel heater cycles. It employs only low pressure feedwater heaters, an open deaerating feedwater heater, and a larger economizer in series with the feedwater heaters. The temperature entropy diagram for the exhaust-fired system is illustrated by Fig. 76. A typical schematic system is illustrated by Fig. 77.

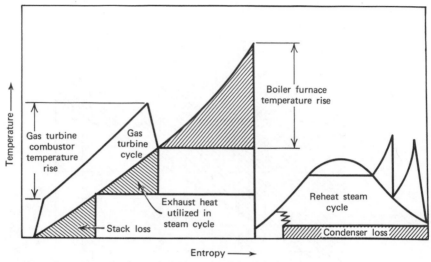

Fig. 76. Temperature entropy diagram for exhaust-fired combined-cycle system.

The net heat rate for an exhaust-fired system is between 7900 and 8250 Btu/kWh, with the steam turbine carrying 60% of the total load. Heat rates rise to between 8200 and 8400 Btu/kWh at 20% steam capability. Exhaust-fired heat recovery systems have a standby forced draft fan to permit cold air firing during inspection and overhaul of the gas turbine. Bypass stacks and dampers are usually included so that the gas turbine can operate independently of the steam system for peaking duty during steam turbine and boiler inspection and overhaul. Figure 78 illustrates typical installation costs for exhaust-fired heat recovery systems.

Currently available combined-cycle power plants are restricted to liquid and gaseous fuels, because gas turbines require these fuels. Suitable liquid fuels for gas turbines are ASTM GT–2, GT–3, naptha, and crude or residual oils low in ash, particularly vanadium and sodium. The supplemental-fired heat recovery boilers generally require high quality fuel that is low in ash because of the difficulty in cleaning the ash deposits from the finned tubes.

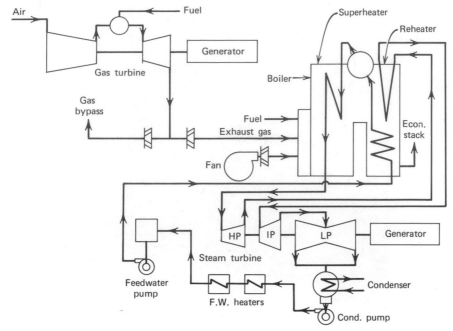

Fig. 77. Schematic for exhaust-fired combined-cycle system.

The boiler in the exhaust-fired cycle can be fired with heavy oils. Therefore, the exhaust-fired cycle offers the greatest fuel flexibility of all combined-cycle plants. It also permits firing of crude oil by a small distillation unit for separating light distillates for firing in the gas turbine and burning the heavier ends in the boiler. The major advantage of combined-cycle power plants in terms of environmental considerations is their low heat rejection to the condenser, which reduces cooling water requirements. If air cooling is required, the cost of cooling towers or air-cooled condensers will be the lowest for any major thermal power generation system. Stack emissions are usually not a difficult problem because combined-cycle plants require high quality fuel. Water injection systems can be used to limit NO_x emissions to comply with 1975 requirements of the Environmental Protection Agency. The fired boiler combined-cycle systems form a two-step combustion process, with one step in the gas turbine combustor and the other in the boiler furnace. This arrangement reduces maximum gas temperature in the boiler furnace, which tends to reduce NO_x formation.

Steam injection. Steam injection to a gas turbine combustor as a source of incremental power has been given some study. Steam for injection is

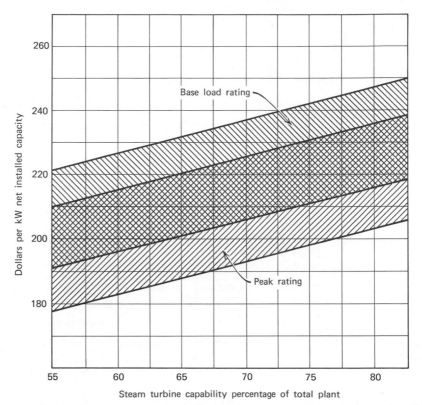

Fig. 78. Installation costs for exhaust-fired recovery cycle. Costs projected to 1975 estimates.

generated by recovery of gas turbine exhaust heat. Steam is injected with some superheat directly into the downstream portion of the combustion chamber. Upon entering the chamber, the steam is further superheated and in the process cools the combustion gases. One immediate result is the reduction in exhaust temperature. Since gas turbine capability is a direct function of exhaust temperature, horsepower can be increased by increasing the load until the maximum continuous temperature is again reached.

Some designers have proposed that a steam injection rate of up to 3% of the mass flow rate of air could be tolerated. This results in a total power improvement capability of 10%. Tests have indicated that the water rate for the steam injection increment is 17.7 lb/(hp)(h), which is not as good as can be attained by a separate condensing turbine. Further development may make possible at least 20% additional power from steam injection. It is further expected that steam injection can be increased to the equivalent of a low ambient temperature mechanical limitation on a 95°F day.

Steam injection is a comparatively new method, and there are as yet little data on it. It is a relatively simple method and could be quite acceptable for small power increments, generating the required power with no additional fuel expenditure. The boiler can be simple, either an unfired heat recovery boiler or a supplementary-fired heat recovery boiler. Another mode is generation of steam at 650 psig, 750°F to drive a back-pressure steam turbine for additional power and then exhaust or inject the steam at about 150 psig into the gas turbine combustor.

One basic disadvantage is that all the condensate is lost with the gas turbine exhaust gas. This requires larger feedwater treatment equipment, increasing the cost of boiler water. Another disadvantage is that more power per pound of steam can be realized by the use of condensing turbines in heat recovery combined cycles, as described previously.

Municipal Waste Incineration

The generation of high pressure steam by heat recovery from the incineration of municipal wastes has had wide acceptance in Europe but is only now being considered by some cities in the United States.

This subject is not strictly concerned with waste heat recovery because the boilers used are somewhat similar to conventional-fired boilers, except that the combustion chamber and fuel feed system are specifically designed to handle solid waste material from residential and/or industrial sources. The refuse fire chambers, including accessory equipment, differ fundamentally from conventional plants in design and operation.

The core of every refuse incinerator is the firing chamber. Incineration must be adapted to a wide range of varying refuse properties, in contrast to conventional furnaces operated with one or several strictly specified fuels. In some areas of Europe, a lower heating value of 1440 to 4500 Btu/lb is demanded for refuse by the consumer. Variations in moisture content of up to 50%, in ash content of up to 60%, and in combustibles of at least 25%, for example, are agreed upon for incineration without auxiliary heating. Each fire chamber has a maximum and minimum incineration capacity in tons per hour, depending on its grate area. The limiting output depends on the combustible material in the refuse and the composition of the refuse (primarily cellulose or mainly carbon). In addition, incineration capacity is limited by the upper and lower refuse heat output in Btu/h, which results from the thermal load on the grate area in Btu/(sq ft)(h).

On the boiler side, heat supply limits from refuse incineration are established by the maximum permissible superheater and waste gas temperature on which engineering and choice of materials are based. Knowing the gas mass flow in pounds per hour and gas temperature out of the combustion section, one can calculate the heating surfaces using methods discussed in previous chapters. The pressure section of the boiler has to be constructed for a fuel with a very high ash content and must be provided with sufficient cleaning facilities. The operator of a boiler combined with refuse incineration must deal with special conditions with regard to operating time and corrosion on the pressure section. These operational problems—such as short operating time of $1\frac{1}{2}$ to 3 months and corrosion on certain heating surfaces— can be reduced somewhat by suitable preparation of refuse to be used as fuel. In case of wide fluctuations in refuse composition, good mixing is necessary. Complete combustion can take place only by a homogeneous distribution of the combustion air.

This is required to maintain fouling and corrosion of heating surfaces within tolerable limits. The corrosion rate is partly a function of fouling; thus its reduction increases boiler life and extends operating time.

Fouling on the flue gas side is the main problem in most large-scale plants. After 1200 to 2000 hours of operation, the plant must be shut down, because extensive fouling of heating surfaces increases draft loss and increases stack temperature. Fouling depends on fly ash content and composition as well as tube section configuration. Slag formation is considerably greater on horizontal than vertical tubes, and vertical or suspended heating surfaces are easier to clean. Obviously, boilers constructed with suspended heating surfaces are also easier to clean. Boilers constructed with suspended vertical tube banks must operate on the forced recirculation cycle. Moreover, all tubes must be bare, as extended surfaces foul almost immediately. The slag consists in part of low melting eutectics, the formation of which is explained by the refuse composition. Therefore, the deposits remain pasty and tacky down to temperatures of 1025°F and thus will adhere up to the area of the pre-superheater. Nearly all refuse incinerators are operated with an excess air rate of 70 to 80%, and reduction processes must therefore take place within the slag deposits. A local ignition caused by glow-fly ash constituents can often be observed on fouling deposits of superheater tubes; this explains the formation of sulfides. In regular control of wall thickness, material wear approaches a limiting value, which allows the anticipation of an acceptable tube life. The reason for this is a certain corrosion protection by the slag formation on the heating surfaces. Therefore, during cleaning of the tubes, care must be taken that a hard shell of 0.12 to 0.20 in. is left for basic fouling as a corrosion inhibitor. The corrosion rate increases almost linearly up to about 1000 hours of operation and then flattens out gradually, approaching a limiting value asymptotically. Figure 79 illustrates this action. A listing of absolute values is omitted because they differ for every plant and depend on

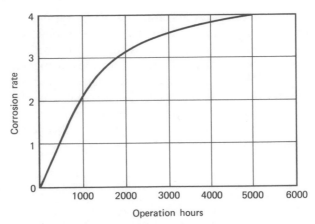

Fig. 79. Corrosion rate versus time.

refuse composition, operating procedure, and dilution of refuse flue gases (i.e., whether refuse incineration is the main firing contributor in the boiler or only a secondary contributor). In refuse incineration with a large percentage of coal ash during the winter months, the corrosion rate decreases compared with summer operation. Ash from coal furnaces inhibits corrosion. This has been confirmed in combination plants in which powdered coal firing is the main fuel through which refuse gases must pass.

The erosion of tubes at certain points in the firing chamber has a similar appearance to that on the final superheater (i.e., layers of scale that easily flake off are located under the shell-like deposits). While the process slows down at the superheaters, the corrosion rate on the heating surfaces of the fire chamber continues approximately linearly because of the flue composition in the influence of secondary air. Channeling of the atmosphere in the furnace with zones of air deficiency, and thus reduction processes in the combustion chamber, cannot be avoided. To this is added the influence of the flame front, which contains a large quantity of glowing, and therefore incompletely burned, parts that continuously subject the iron oxide layers to renewed reduction on contact. It is evident from the foregoing considerations that high temperature corrosion is involved in all cases.

Several theories exist concerning the chemical process of corrosion. The refuse or fouling deposits, respectively, contain practically all elements of the periodic system. Consequently, the most diverse corrosion processes can take place among which the formation of complex alkali sulfates and of chlorides together with reduction processes probably constitute the major contribution.

The wide acceptance of polyvinyl chlorides (PVC) has had a marked effect on the design of incinerators that burn PVC waste products and hence on heat recovery equipment that operates in conjunction with such furnaces. Even if it were possible to interpret completely, the chemical mechanism of corrosion in a refuse burning facility, it would hardly be possible to derive suitable countermeasures. This points up the importance of careful planning in the design of a refuse burning and heat recovery facility. The following points are derived from experience in the design and operation of refuse burning and heat recovery equipment.

1. *Fire chamber side.* Adjustable bed height over the charging grate, resulting in improved metering corresponding to refuse constituents. Increase of the flow resistance in the grate bed unifying air distribution over the undergrate zones, independent of the fuel bed. Arrangement of a fire bridge for which an oil ignition burner creating effective secondary air supply for homogenizing the flue gas atmosphere and for complete combustion. Air excess of 80%, fluctuating between 60 and 100%.

2. *Boiler side.* Ceramic combustion chamber in the entire flame zone, with only a grate cooling band to prevent caking and damage on the chamber wall in this region.

   Radiant heating surfaces as smooth side walls consisting of evaporator tubes should be used.

   Superheaters must be located outside the refuse combustion zone. Refuse flue gases must be thoroughly mixed with the total flue gas stream before the air preheater.

   Total heating surface should exceed normal design requirements by about 30% to allow for the fouling effect of refuse fly ash.

   The boiler should be designed so that heating surfaces are accessible for inspection and cleaning. Pendant-type convection sections lend themselves particularly well because they can be designed to be removable through the top of the boiler in sections. See Fig. 80 for a typical forced recirculation refuse firing boiler arrangement. Heat recovery boilers for refuse incinerators must be equipped with effective soot blowers strategically placed to cover all convection surfaces.

OTHER FORMS OF HEAT RECOVERY BY STEAM GENERATION

Other heat recovery possibilities for various processes that will not be covered in detail here are listed below.

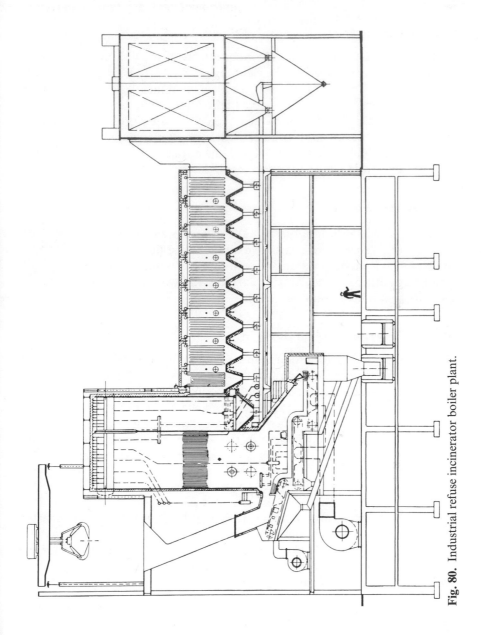

Fig. 80. Industrial refuse incinerator boiler plant.

H₂S Combustion

H_2S is fired into a combustion chamber attached to a fire tube boiler. The fire tube arrangement makes tube cleaning easy—a process required in such an application. The fire tube boiler has a conventional design except that it must have tubes that can withstand high sulfur conditions. Larger sulfur burning boilers are of the water-tube type. Corrosion problems are eliminated by the absence of absorbing brickwork. Boiler refractory setting is replaced by circular, membrane-wall type heat transfer surface, with external gas-tight structural plate casing and insulation.

Glass Melting Furnace

Boilers for glass melting furnaces have a large water wall radiant surface with succeeding passes of convective surface. The final section is usually a combustion air preheater. Furnace walls are gas-tight, with membrane-wall type heat transfer surface. Heating surfaces and walls are designed to minimize the possibility of dust collection.

Oxygen Converter Cooling Hood

Heat recovery from steel production cooling hoods is in the form of forced recirculation steam generation for all steam pressures. Equipment for this application is available for fume cooling only, or as a heat recovery boiler for fullest utilization of the sensible heat in the waste gases. Accumulators and separately fired steam superheaters are also built for use on converter hoods. Dust collection equipment in the form of electrostatic precipitators or wet scrubbers is used after the gas cooling heat recovery equipment.

Corner Tube Type Waste Heat Boiler

Heat recovery boilers for this application must be designed for cooling gases that have an exceptionally high dust burden. Boilers are usually of the forced recirculation type composed of multiples of convection sections cleaned by recirculating shot. The boiler must be designed for ready inspection and accessible for service and maintenance.

Petrochemical Cracking Furnace

Boilers for this service are usually built like a shell-and-tube heat exchanger with controllable central gas bypass duct. They are designed for use on cracking furnaces where gaseous products are at high pressure. Tubes must be arranged to accommodate differential expansion. Boiler design for this application uses data for gases in turbulent flow inside tubes and water in natural convection outside tubes.

Ammonia Combustion

Boilers for this application may use either natural or forced circulation. Corrosive conditions require attention to selection of tube material.

Zinc Recovery

Boilers for this application generally feature membrane walls at the gas entry section to cool carryover below softening point. Membrane walls, welded tube-to-tube, are not subject to fouling. Hairpin or pendant evaporator and superheater tubes must be generously spaced to avoid fouling. There should be numerous access doors and cleaning lanes along the entire length of the boiler. Ash hoppers are provided under the pendant tube assemblies, and vibrators are sometimes used to shake deposits from the tubes and into the hoppers. Because of the pendant coil construction, forced recirculation must be used.

In the design of any kind of boiler for the uses described above, the appropriate curves presented earlier for gas and liquid heat transfer and pressure drop are to be used with particular attention to environmental considerations such as corrosion, erosion, dust accumulation, and the like.

HEAT RECOVERY IN DESALTING APPLICATIONS

Heat recovery from engine or gas turbine exhaust is used as low pressure steam to provide the motive steam for a multiple-effect desalting system. The shaft power generates electricity to drive pumps, compressors, and so on. The design principles of such heat recovery equipment are the same as previously described under low pressure steam generation from engine and gas turbine exhaust.

For efficient and economical application of the gas turbine or engine, there must be a requirement for power developed at the shaft and for exhaust heat. Both shaft power and exhaust heat can be utilized in gas turbine desalination processes for water.

Several process flow schemes are available to engineers for incorporating the gas turbine or engine in both the vapor compression and multi-stage flash-type distillation processes. Combined electric power generation and waster desalting are also possible.

Enormous quantities of energy are required to operate the large desalting plants now planned or under construction. Both heat and mechanical energy are required. Studies are under way to combine nuclear power and large-scale desalting. The proposed capacity is 150 million gal per day plus 1800 MW of power.

On the other end of the scale are the small to medium-sized plants with capacities of 10,000 to 2 million gal per day. In this category are water plants for hotels, resorts, industrial complexes, military bases, small municipalities, bottled water plants, food product manufacturing, and steam electric generating plants. Although the large plants will continue to receive the major attention, the need for the smaller plants will continue to expand at a faster rate than in the past. For this requirement, the gas turbine will find increasing use when properly applied.

Gas turbines are used in all three areas of electric power generation: (1) base load, (2) peaking, and (3) stand-by operation. Since a combined electric power and water plant must be in continuous operation, its ability to produce power and useful heat make its application acceptable. With a fully fired heat recovery boiler backing up the gas turbine generator, overall fuel efficiency can range up to 80%. A 750-kW gas turbine generator can supply sufficient heat energy from its fired heat recovery boiler to produce potable water from sea water at the rate of 150,000 gal per day, sufficient for a community of about 3700 people.

The vapor compression cycle uses both shaft power and exhaust heat from the gas turbine. This cycle is widely accepted for desalting and operates at efficiencies higher than those of evaporation methods.

To minimize the size and cost of the steam compressor, the vapor compression system should be operated just above atmospheric pressure—as with a corresponding vapor inlet temperature to the compressor of about 215°F and an exit saturation temperature of about 225 to 235°F. At these temperatures, sea water and most well waters deposit scale very rapidly, which requires a shutdown for acid cleaning every 20 to 30 days. This problem has been overcome, however, by continuous acid treatment of the incoming feedwater followed by deaeration and decarbonation, with the result that modern vapor compression units operate scale free and without corrosion. This development is expected to increase significantly the use of the vapor compression cycle. Very small units use a diesel engine, an electric motor, or a gasoline engine for driving the compressor. Large plants, in the range of 500,000 to 1,000,000 gpd, will use the compressor driven by gas turbine with the efficient use of the exhaust heat.

One method of utilizing exhaust heat from the gas turbine is recovery in the form of steam, with operation of a secondary flash evaporator. This system

system offers the designer great flexibility in meeting both power and water requirements.

FIRE TUBE-PROCESS GAS BOILERS

When high pressure gases are involved, fire tube boilers (in which gas passes on the inside of the tube and water on the outside) are necessary. In the steam-methane reforming process, for example, the high pressure process gas leaves the reformer at 1600 to 1800°F and several hundred psi.

Because the gas must be cooled before it enters the next step of the process, and because steam is needed by the process itself, cooling should be done through a boiler. But since the casing of a water tube boiler cannot stand the gas pressure, a heavy pressure vessel is required. A simpler solution is making the gas flow inside the tubes, using tubes capable of withstanding the pressure.

Often the hot gas must be controlled to a certain temperature for further processing. A gas bypass with butterfly damper can regulate the gas outlet temperature at any load condition. This damper can be controlled manually or automatically, as required by the process.

In most designs, the bypass is internal, permitting a compact design. Gas temperature in the boiler can be precisely controlled with this device using dampers. Thermal expansion differential between the fired tubes and the shell and between the tubes and the internal bypass pipe can be a problem, however.

Tube sheets are the most vulnerable parts of fire tube boilers. Tube joints, leaks or even cracks in the tubesheets often result from faulty design. In such a heat exchanger, the temperature of both the cooled and the cooling medium changes along the heat exchange surface, and the heat transfer coefficient might be the same magnitude on both sides of the tube. This means that the temperature of the tubes is considerably higher than the temperature of the shell wall, which is not exposed to heat transfer. The resulting thermal expansion difference, without provisions such as an expansion joint on the shell or a floating tube sheet, creates stresses resulting in damage.

In a fire tube boiler, the steam-water mixture is at the same temperature (saturation) in every part of the shell. In addition, the film coefficient on the water side is 30 to 300 times that of the gas side. Thus both the tube and the shell are at about the same temperature. Any thermal expansion stresses are slight and can easily be sustained without damage by the shell, the tubes, and the tubesheets.

The hot tubesheet, the most vulnerable part of the fire tube waste heat boiler, requires special attention. When the inlet gas temperature exceeds 1400°F, it is customary to protect the tubesheet from the radiant heat of the gas with a layer of refractory. Stainless steel, Incoloy, or ceramic ferrules inserted into the inlet of each fire tube prevent refractory particles from being carried into the tubes by the hot gas stream. These ferrules also help remove the maximum heat flux from the tubesheet as a result of the air space between tube and ferrule. This prevents high temperature differences around the circumference of the tubesheet. When calculating the required thickness of the refractory lining, keep in mind that the "K" factor of the refractory is higher in atmospheres containing hydrogen and increases as a function of the percentage of hydrogen present. A low silica refractory should be used with a high temperature gas stream; otherwise the silica might evaporate and deposit on the cold tube surface, lowering the heat transfer rate and causing excessive gas outlet temperatures.

Heat flux is a most important design parameter for process gas boilers. The high pressure and the low specific volume of the gas permit a high mass flow without excessive gas velocity or pressure drop. Therefore, the gas-side heat transfer coefficient is usually high, around 100 Btu/(h)(sq ft)(F°). With a high gas inlet temperature, such rates can produce a high heat flux at the tube inlet and cause film boiling, which leads to overheating and loosening of the tube joint.

Film boiling and burnout are complex phenomena. Not the direct function of heat flux alone, they also depend on temperature differences, water chemistry, tube diameter, and circulation. At ideal conditions, film boiling might be expected to begin at heat fluxes in the range of 400,000 Btu/(h)(sq ft).

In a well-designed process heat recovery boiler, therefore, maximum heat flux should be 100,000 Btu/(h)(sq ft). This limit offers adequate protection against the possibility of burnout. It is not good design practice to set a limit for "average heat flux," because in units with a low approach temperature (long boiler), the maximum heat flux at the tube inlet could exceed the limit even if the average heat flux is below the specified limit. However, in boilers with a high gas outlet temperature, average heat flux may be well above the arbitrarily chosen average, while the maximum can still be below the dangerous upper limit.

Circulation is the most important design consideration. Most boiler failures can be traced to inadequate circulation. Fire tube boilers can be designed with an integral steam space but with a separate steam drum. This arrangement with its higher differential head assures better circulation. There are added advantages with a separate drum: (1) all tubes will always be submerged in water; (2) steam quality will be improved; and (3) both tube sheets will be exposed to the same heat transfer conditions, not to steam at the top and water at the bottom, as with an integral steam drum.

With a separate drum, sizing of risers and downcomers related to the available static head determines the circulation ratio. The correct location of the risers and downcomers ensures effective circulation in each part of the boiler and prevents dead spaces.

In a single-pass fire tube heat recovery boiler, circulation parallel to the shell is preferred. To achieve this, the riser tubes should be close enough to the hot tubesheet so that all generated steam can be immediately released to the steam drum, preventing steam blanketing of the hot tubesheet. This design exposes the tubesheets to similar steam-water mixtures, creating even heat transfer conditions. In addition, the heavy turbulence keeps the hot tubesheet well cooled. Circulation is completed by reintroducing boiler water to the cold end.

Circulation perpendicular to the shell might also be effective if both risers and downcomers are closely spaced along the shell. In this case, the bottom of the hot tubesheet is exposed to water only, and the top to a mixture with maximum steam content. The different heat transfer conditions might impose additional stress on the tubesheets.

In the first case, a single blow down nozzle at the bottom of the shell close to the hot tubesheet is very effective, while in the second case, several nozzles are needed because the mud is not carried to one location by the circulation.

WATER TUBE-PROCESS GAS BOILERS

The trend in ammonia and methanol plants is toward steam generation at pressures above 1000 psig. The fire tube boiler is not suitable for such service; therefore, water tube boilers have been developed. In this type of boiler, the high pressure gas flows through a pressure vessel containing serpentine tube bundles that absorb the heat from the high pressure gas. The tube bundles can be designed for either forced or natural circulation. In either case, an external steam drum is required. The heating surfaces are removable through a flange at one end of the pressure vessel.

WASTE HEAT RECOVERY IN STEELMAKING

Utilization of the heat in exhaust gases from industrial process furnaces becomes more important as fuel costs increase. Boiler equipment properly designed to absorb the heat in what was formerly waste gas often generate all the steam required to power the process. Where the waste gases carry some of the process material in suspension, suitable hoppers associated with the boiler equipment collect a portion of the material, and the cooled gases leaving the boiler may be passed through precipitators for a major recovery of the remainder. Many types of boilers are necessary to meet the diverse requirements in this field. Their design depends on the chemical nature of the gases, their temperature, pressure, quantity, and dust loading.

The rate of heat transfer from gas to boiler water depends on the temperature and specific heat of the gases, their velocity and direction of flow over absorbing surfaces, and surface cleanliness. To obtain the proper velocity of gases over the surfaces, sufficient draft must be provided, either by a stack or a fan, to overcome losses from gas flow through the unit, with adequate allowance for normal fouling of heating surfaces. The temperatures of many process gases are relatively low, as shown in Table 11. The radiation component in heat transfer in industrial off-gas applications is low, with a consequent tendency in the design of many waste heat boilers to use higher gas velocities than prevail on fuel fired units. However, high velocities with dust laden gases must be avoided to prevent abrasion of tubes. This factor is particularly critical where changes gas flow occur.

TABLE 11 Temperature of Waste Gases

Gas Source	Temperature (F°)
Ammonia oxidation process	1350–1475
Annealing furnace	1100–2000
Cement kiln (dry process)	1150–1500
Cement kiln (wet process)	800–1100
Copper reverberatory furnace	2000–2500
Diesel engine exhaust	1000–1200
Forge and billet-heating furnace	1700–2200
Gas turbine exhaust	800–1000
Garbage incinerator	1550–2000
Open-hearth steel furnace, air blown	1000–1300
Open-hearth steel furnace, oxygen blown	1300–2100
Basic oxygen furnace	3000–3500
Sewage sludge incinerator	1000–1400
Sulfur ore processing	1600–1900
Glass melting furnace	1200–1600
Zinc fuming furnace	1800–2000

Tables 12 and 13 give an approximate measure of the convection heating surface required for usual conditions in heat recovery boiler practice. A water-cooled "vestibule," or furnace, is a feature of many modern waste heat boiler units. The vestibule cools the gases to the temperature necessary for prevention of slagging before they enter the convection surface section. The approximate amount of surface required for this purpose is given in Table 14.

TABLE 12   Convection Surface vs. Gas Temperature Entering, at W/A = 2000 lb/(sq ft)(h)

Gas Temperature (F°)	Convection Surface, sq ft/1000 lb of gas at stack temperature (F°)		
	550	650	750
1000	80	45	27
1200	90	60	40
1400	105	68	50
1600	115	77	60
1800	122	82	65
2000	124	90	70
2200		92	73
2400		95	76

TABLE 13   Convection Surface vs. Gas Temperature Entering, at W/A = 3000 lb/(sq ft)(h)

Gas Temperature (F°)	Convection Surface, sq ft/1000 lb of gas at stack temperature (F°)		
	550	650	750
1000	62	35	20
1200	73	47	32
1400	80	55	40
1600	83	63	45
1800	90	67	50
2000	97	70	55
2200	102	75	60
2400		77	63

TABLE 14   Vestibule Heating Surface vs. Gas Temperature Entering, at saturation steam temperature of 450°F

Entering Temperature (F°)	Vestibule Surface, sq ft/1000 lb of gas at steam temperature of 450°F and leaving temperatures				
	1200	1400	1600	1800	2000
1200					
1400	14.0				
1600	21.5	9.5			
1800	29.5	12.5	7.0		
2000	32.0	20.0	11.0	6.0	
2200	39.0	22.0	13.0	10.0	4.5
2400		28.5	19.5	11.5	8.0

The modern basic oxygen steel-making furnace is blown with pure oxygen through a retractable water-cooled lance mounted vertically above the furnace. After each charge of molten iron, scrap steel and fluxing material is loaded into the furnace, the oxygen lance is lowered into position above the charged material, and a blowing period of 15 to 20 min commences. During this period, the oxygen starts a chemical reaction that brings the charge up to temperature by burning out the silicon and phosphorus impurities and by reducing the carbon content, as required for high grade steel. Large amounts of carbon monoxide gas at 3000 to 3500°F are released in this conversion process. The gas is collected in a water-cooled hood and burned with air introduced at the mouth of the hood. The combustion products are subsequently cooled by the addition of excess air, the injection of spray water, or water cooling of the hood. Combinations of these methods may be used. The flue gas finally enters a clean up system for electrostatic or wet scrubbing for the removal of dust particles before discharge to the atmosphere. The service demands limit the life of furnace linings to only a few weeks. Therefore, two or three furnaces, with sequential operation and relining, are usually installed for continuous steel production.

The high temperature (3000 to 3500°F) of the gases discharged from the basic oxygen furnace and their high carbon monoxide content (about 70% by volume) make them ideal for burning in the hood. While there are basic similarities to usual boiler service, there are also significant differences, particularly the carryover of iron bearing slag from the furnace and the short, intermittent operating periods. Accordingly, the criteria established for good hood design are listed as follows:

1. Adequate structural strength. The service is severe, and the equipment is roughly handled.
2. The hood surface in contact with the furnace gases must be smooth to shed more readily the skulls or slag that is heavy with iron.
3. There should be a minimum of cracks, crevices, and sharp corners, and there should also be openings in the fore part of the hood to permit the anchorage of slag.
4. Positive and uniform water cooling of all surfaces exposed to the hot gas must be provided. There should be minimal temperature differences between all water circuits, with no eddy currents or uncooled corners.
5. The hood water walls should be cooled with treated and deaerated water to prevent internal deposits of hard scale or oxygen corrosion. Good water treatment is important because of the high rate of heat absorption.
6. Water circulation should be maintained at a high rate through the entire cycle. Recirculation makes possible high flow rates without excessive makeup.

7. The hood cooling system should be suitable for pressurizing to permit steam generation or high temperature water, in keeping with industrial boiler practice.

The membrane wall for oxygen furnace hoods should ideally meet these design requirements. The membrane wall can be formed into a variety of hood configurations to suit plant layout. The hood may be a long flue type to transport the gases to an evaporating or quench chamber, or it may be a bonnet type that collects the gases and immediately discharges them into a spark box where the temperature is sufficiently reduced with spray water for use in the gas clean up system. Selection and arrangement of the auxiliary equipment may well provide other arrangements with equally good results. Gas temperature reduction with spray water and use of a wet scrubber clean up system or an electrostatic precipitator affect the shape of the hood. The hood with water-cooled membrane walls may be applied to the oxygen converter process in the following alternative ways:

1. It may be operated as a boiler at pressures from 100 to 1500 psi to generate steam for general plant use.
2. It may generate steam that is condensed in a closed system with the heat dissipated by an air-cooled heat exchanger.
3. Water in a closed system may be heated in the membrane walls of the hood and the heat dissipated by an air-cooled heat exchanger.

The oxygen converter hood, when equipped with a steam drum, boiler circulating pumps, boiler mountings, and controls, generates steam effectively during the oxygen blowing period of the converter cycle. Steam generation is limited to the oxygen blowing period because of the intermittent or cyclic operation of the steel-making process.

The rate of steam generation varies from zero to a maximum and back to zero during the oxygen blowing period—normally about 20 min for a complete cycle of 40 to 50 min. This cyclic operation, coupled with the outage time for relining the converter vessel every second week, limits the steam production of a single hood to 12 to 15% of the life of the lining.

Steam during generation may be discharged directly into the plant's steam mains. The cyclic type of operation, combined with short-period high rates of generation, imposes widely fluctuating load swings with which the steam system must cope. The effect of load swings can be reduced by operating the hood boiler unit at a higher pressure and discharging the steam into an accumulator. Heat is thus stored in water at saturation temperature by the rising accumulator pressure. When the steam production rate in the boiler hood drops, the heat stored in the high temperature water of the accumulator is released to produce steam at the lower pressure of the plant's steam mains.

Some blown oxygen furnace plants cannot presently utilize the potential steam production of hood boilers. However, if the use of steam can be anticipated, the units can be arranged for closed-circuit operation. The closed-circuit system assures an ample supply of good boiler water without an elaborate treating plant. A portion of the heat absorbed during the blowing period raises the system pressure to between 250 and 450 psig. Any excess heat is discharged to the atmosphere through an air-cooled condenser operating at system pressure. The collected condensate is returned to the hot well and from there to the hood drum, thereby completing the cycle.

The air-cooled condenser of the closed pressurized circuit is physically small because of the high temperature difference (about 350°F) between the condensing steam and the cooling air. The power required for dissipating the heat is small compared to the pump power for an equivalent system using cooling water as the condensing medium. The power required for water circulation is also small. Makeup is limited to the pump packing leakage and leakage from valve stems.

This closed system can then be readily altered to supply plant steam. All that is necessary is to tap into the steam line from the hood. Steam can be taken from the hood drum, and the load on the air-cooled condenser is reduced by the amount of steam withdrawn.

Some steel mills find it advantageous to forego recovery of heat absorbed by hoods. For these plants, the steam-pressurized, high temperature water, closed-circuit system is preferred. This installation is simpler to control and less costly than the equivalent closed-circuit generating system.

The steam-pressurized, high temperature water system serves the same purpose as the closed-circuit steam system. The main difference is that saturated temperature water is generated in the hood and discharged into the water system's steam-pressurized expansion tank. The saturation temperature water from the expansion tank is pumped through the air-cooled heat exchanger to lower its temperature and to return it to the hood, completing the circuit. With this system, the high temperature water is pressurized to between 250 and 450 psig by controlling the air flow over the heat exchanger.

Oxygen blowing of the open-hearth furnace, in preference to air, greatly increases the production rate and significantly reduces the cost of steel. Therefore, most open-hearth furnaces of recent design are being changed to the oxygen blown process.

Oxygen blowing, however, increases dust discharge. In addition, gas is discharged at higher temperature (as high as 2100°F). The combination of increased dust carryover, higher gas temperature, and, in many cases, increased total gas flow makes it desirable to replace the gas tube boiler with a water tube unit of higher capacity. Furthermore, because of the increase in dust carryover, the new boiler is usually arranged with a new and larger precipitator.

As a result of the increase in both gas weight and temperature, steam output is increased appreciably. For many installations, it is desirable to maintain steam flow during furnace charging and maintenance periods by firing an auxiliary fuel, which requires a boiler furnace for combustion. Thus, the waste heat boiler for the oxygen blown open-hearth furnace is a versatile unit that takes into account available space, waste gas quantity, steam capacity, cleanability, and supplementary fuel firing. The steam capacity for a single unit serving a single open-hearth furnace may be as high as 150,000 lb/h.

FORCED RECIRCULATION

This form of steam generation for heat recovery applications deserves separate comment because there are instances, as cited in the foregoing discussion, that demand forced recirculation.

Reliability and durability are gained by the use of forced recirculation systems, which prevent vapor separation within the steam generating tubes. The rate of recirculation is usually 5 to 10 times the maximum steaming rate of the boiler. Forced recirculation makes possible the use of smaller-sized tubes, which reduce boiler size and weight and reduce the time required for steam up. Only one drum is required. Some forced recirculation boilers, particularly smaller ones, have only a cyclone steam separator with an enlarged lower section for water level control, instead of a steam drum. Boilers having only a cyclone separator are particularly adaptable to shipboard use because they are not particularly sensitive to pitch and roll.

The primary advantage of forced recirculation is probably the assurance that there is always forced water circulation in all boiler tubes. Since heat recovery boilers very often have limited space, there may not be sufficient distance between the upper and lower drums to promote effective natural circulation. Forced recirculation is usually the solution in such circumstances.

A forced recirculation boiler is nearly always more compact than a comparable natural circulation boiler, since the former does not require a fixed relationship between the drum and the heating surfaces. The drum or separator can be located almost anywhere adjacent to the heating surface assembly.

Heating surface design can be quite flexible. Vertical or inclined tubes are not required. The heating surfaces can be arranged in a horizontal return bend configuration or a circular shape consisting of flat, spirally formed coils or other shapes. It is always preferable to have a drainable coil assembly.

External clean out plugs permit internal cleaning of individual tube circuits either chemically or mechanically, without disassembling the boiler casing. Cleaning of external surfaces of the tubes is by conventional soot blowers, operated either by air or steam.

No internal baffles are required in forced recirculation. Internal baffles are difficult to install and maintain and can cause ash accumulation unless they are steeply inclined or vertical. Even inclined baffles require some blowing to keep them clear of soot and ash.

Forced recirculation boilers are generally lighter than comparable natural circulation boilers, requiring less supporting structure and being easier to install.

Probably the most criticized part of a forced recirculation boiler is the recirculating pump. Obviously, if the pump fails, all water circulation ceases in the boiler tubes and the boiler must be immediately removed from service and isolated from its source of hot gas. However, a pump properly selected for service, particularly as to its mechanical shaft seal, will provide long, trouble-free service. The shaft seal and its environment is one of the most critical parts of the system. For operating temperatures above 250°F, it should be water cooled. If the pump is to handle water containing any foreign matter, the seal faces must be flushed. It is advisable to install a duplex strainer ahead of the pump to eliminate larger particles. In some critical applications, a stand-by pump is included. Another important item in the proper selection of a pump is NPSH. This could be important on close coupled systems with only a few feet of head on the pump suction. Failure to consider this factor could result in severe cavitation and consequent re-circulation failure. Recirculation pumps for capacities up to about 30 gpm may be regenerative turbine pumps. These pumps are particularly sensitive to dirt because of the close running clearances between the casing and the impeller. For flows in excess of 30 gpm, a single-stage centrifugal pump is used. The differential head on a typical recirculating pump can vary from 50 to 75 ft, depending on pressure drop through the tubes, tube inlets, and steam separator nozzle, if used. For best results, the pump should be operated as close as possible to the design points on its performance curve.

Water and steam velocities in the tubes of a forced recirculation boiler vary with load and pressure. As a general rule, the recirculating pump is sized to produce a water velocity of 2 to 4 fps at the entrance to the tubes where the fluid is all water. Velocity increases as heat is absorbed by the water and steam is generated. At the point where the water-steam mixture discharges into the steam drum or separator, there will be approximately 10% steam and 90% water when the boiler is at full load and the recirculation rate is 10 to 1.

The selection of tube diameter depends on boiler size, configuration, and end use. Tube sizes may vary from $\frac{3}{4}$- to 2-in. o.d. Small, high performance boilers have smaller tubes. However, in making this choice, feedwater quality must be considered. It is advisable to provide for individual tube circuit cleaning by having removable plugs opposite each tube termination at the inlet and outlet headers. Provision can be made for chemically cleaning any tube size. Mechanical cleaning by a pneumatic tube cleaner is generally restricted to tubes of 1-in. l.d. and larger. The mechanical method of reaming out boiler tubes has proven to be very practical even in bent or circular coil designs having at least an 8-in. minimum radius of curvature.

A forced recirculation heat recovery boiler can be built in a variety of configurations, depending on service requirements. Therefore, it is difficult to specify a typical design. Figure 63 illustrates one schematic of a forced recirculation boiler system having an economizer and a superheater section. Many such systems consist only of an evaporative surface assembly, a recirculating pump, and a steam drum or separator. If a steam drum is used, the same design criteria for a natural circulation system may be used. If a separator is used, it can be either a cyclone or baffle type, although the cyclone produces higher quality steam but requires a higher pressure drop (usually 6 to 10 psi). Figure 81 illustrates a cyclone steam separator design that has been extensively used on forced recirculation steam generators and heat recovery boilers. The water-steam mixture enters the cyclone body through a tangential nozzle, than proceeds downward guided by a helical ribbon. The mixture emerges over a spin plate in the bottom of the separator. A high velocity cyclone is formed on the spin plate, which flings water out toward its periphery, where it is captured by radial vanes that direct the water downward and into the recirculating pump. Steam rises, still spinning to the dry pipe. Water particles are continuously flung toward the wall, where they drain into the pump. The result is steam quality of the order of 99.4 to 99.6%. If higher quality is required, two cyclones can be used in series.

Hydrodynamic or flow instability is a potential problem in forced recirculation evaporators. Instability can result in periodic oscillations in steam drum water level (slugging), steam flow, and steam pressure. Instability may be either static or dynamic. Density wave instability is the most common type of dynamic instability. The effect of various parameters have been studied, and the more pertinent ones are as follows:

1. An inlet restriction increases stability.
2. An exit restriction decreases stability.
3. An increase in two-phase flow pressure drop decreases stability.
4. An increase in steam pressure increases stability.

On forced recirculation systems, inlet subcooling appears to be the most useful single independent variable in determining whether the flow will oscillate. For any particular heat flux, loop geometry, and liquid flow rate, there is a definite range of inlet subcooling within which the flow oscillates. If inlet subcooling is outside this range, the flow is steady.

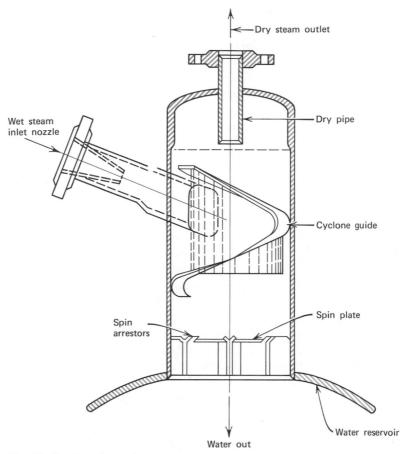

Fig. 81. Section of a cyclone steam separator.

An increase in total dissolved solids in the drum has a destabilizing effect, based on experience with natural circulation designs. A physiochemical hydrodynamic effect and its resultant increase in pressure drop in the separator or drum located at the exit end of the circulation loop is responsible for the destabilization. The solids level must, therefore, be kept at a reasonable level (not over 2000 ppm to prevent carryover).

As stated earlier, an inlet restriction increases stability. Therefore, each evaporator tube in a multi-circuit unit must have a flow restrictor. The restrictors must be located where no boiling occurs and where there is true, single-phase flow. Stable flow can be achieved with a pressure drop for the flow restrictors equal to between 20 and 160% of the two-phase pressure drop. This writer has had considerable success with restrictor pressure drops of approximately 50% of two-phase pressure drop.

In addition to providing a pressure drop, a flow restrictor must fulfill two other requirements. (1) There can be no accumulation of suspended solids at the restrictor or it will plug. (2) The restrictor must be fully drainable, which is important during acid cleaning operations. Therefore, the restrictor should not be a simple thin plate or sharp edged orifice, but a longer, larger-diameter section that produces the required pressure drop but is not susceptible to plugging.

As stated previously, an exit restriction decreases stability. Therefore, the steam separator, which is at the exit end of the evaporation section, should have a minimum pressure drop consistent with efficient separation. It is also important to keep the evaporator section at a low pressure drop. For this reason, larger diameter tubes or multiple parallel circuits are used.

The amount of subcooling entering the evaporator is important to flow stability. Usually, the economizer section is designed so that the water flow leaving it is subcooled between 5 and 10°F. The flow from the economizer is sprayed into the drum above the water level and thereby undergoes additional heating. It then mixes with saturated water being recirculated, so that water entering the evaporator is subcooled by about 3 to 5°F.

Static instability also affects forced recirculation evaporator sections. If the evaporator pressure drop vs. flow curve intersects the recirculating pump characteristic curve, static instability results. The inlet flow restrictor changes the maximum and minimum points to a single point of intersection with the pump curve. Static instability occurs when the slope of the system characteristic curve is algebraically smaller than the slope of the pump curve.

In summary, the potential problem of hydrodynamic instability has been given thorough consideration and features such as inlet flow restrictors, large diameter tubes and subcooling at the evaporator section inlet have been used successfully to produce stable operation.

Steam temperature control in a supplementary-fired heat recovery boiler is accomplished by bypassing some of the saturated steam from the drum around the superheater. This bypassed flow is mixed with steam that has passed through the superheater tubes in the superheater's outlet header. As lower outlet temperatures are desired, more saturated steam is bypassed. The maximum amount that can be bypassed is about 20% of the total.

This type of steam temperature control system has several advantages. It is superior to the system that injects water into superheater steam for cooling because such a system could always send slugs of water into the turbine if the control valve failed. Since steam by-pass control injects dry saturated steam, this problem cannot arise. It is also better than a system that controls steam temperature by varying firing rate because it responds more rapidly to load swings.

Stability in the superheater and economizer can be accomplished by designing these sections for a reasonably high pressure drop: 10 to 20 psi for the superheater and approximately 15 psi for the economizer. Economizer outlet temperature is controlled by constant flow to the economizer, bypassing the portion not required to satisfy the demand diverted back to the deaerator. Steaming in the economizer must be avoided because it causes stagnation in some tubes that begin to steam with an accompanying increase in pressure drop, which, in turn, leads to less flow, more steaming, and ultimately stagnation. With a high pressure drop across the entire economizer in the beginning, the increase in pressure drop caused by steaming in some tubes is small compared to the total, resulting in significant changes in flow and thereby preventing stagnation.

Dry operation of the heat recovery boiler at turbine exhaust temperature is necessary for systems without a bypass. Dry operation permits gas turbine operation whenever the boiler is inoperative. Some designers use carbon steel tubes and fins even for dry operation to 900°F and rate the unit for operation up to 6000 h at 900°F, and can run dry for longer periods at lower temperatures.

In using carbon steels for high temperature dry operation, items to consider are (1) oxidation, (2) graphitization, and (3) creep. Using data on carbon steel, Table 15 was developed.

**TABLE 15   Loss in Thickness Due
to Oxidation**

Temperature (F°)	Loss (mil/yr)
900	0.00
950	3.90
1000	8.70
1050	14.70
1100	24.10
1200	55.90

This table shows that the amount of oxidation at 900°F is zero. A maximum temperature of 950°F for carbon steel in an oxidizing atmosphere has been recommended by tube suppliers. Thus, it may be concluded that 900°F is a safe run-dry temperature. However, other factors must be considered.

Operation with carbon steel at 900°F might cause some concern about graphitization. Refer to Note 1 to Table PG-23.1, Section 1, of the ASME

Boiler Code, which reads: "Upon *prolonged* exposure to temperatures above about 800°F, the carbide phase of carbon steel *may* be converted to graphite." The key words are *prolonged* and *may*. A total of 6000 h of operation at 900°F is not normally considered "prolonged" exposure. Therefore, the steel must be carefully selected. Use only silicon-killed steels and avoid aluminum-killed steels (which are susceptible to graphitization). At temperatures below 900°F, exposures greater than 6000 h can be used without fear of graphitization. At 850°F, 12,000 h are recommended, with indefinite operation at 800°F and lower.

The third consideration is creep. Stresses in affected carbon steel members have been found to be less than the allowable stress of 5000 psi given for carbon steel in Table PG–23.1, Section 1, of the ASME Boiler Code. The stresses are also less than the stress required to produce 0.10% creep in 10,000 h. On the basis of creep, therefore, the internal carbon steel members are capable of operation at 900°F for 6000 h.

The following figures are used in the design of a forced recirculation heat recovery boiler:

Fig. 4—Gas pressure drop versus mass flow
Fig. 6—Approximate mean specific heat of flue gas
Fig. 18—Base value of film coefficient for gases outside tubes
Fig. 19—Temperature correction factor for Fig. 18
Fig. 38—Gas pressure drop versus mass flow, gas outside finned coils
Fig. 39—Gas pressure drop through portions of heat recovery unit
Fig. 40—Base value of film coefficient for liquids inside tubes, turbulent flow
Fig. 41—Correction factors for Fig. 40
Fig. 42—Fin efficiency chart
Fig. 43—Water pressure drop in coiled pipes
Fig. 44—Loss of head from sudden enlargement
Fig. 45—Loss of head from sudden contraction

The first design item to decide is maximum allowable gas pressure drop. With this factor, gas mass flow rate G in lb/(sec)(sq ft) can be found from Figs. 4, 38, and 39. Sometimes a second trial must be made.

If we know the total recirculation rate (10 times the expected maximum steam rate), service conditions, and proposed configuration, we can select the appropriate tube size.

Gas mass flow rate and tube diameter are used in Figs. 18 and 19 to determine the gas-side film coefficient of heat transfer.

Using Fig. 43 for pressure drop, and assuming that the maximum allowable pressure drop is known (usually 12 to 20 psig through tubes only), with an initial velocity of 2 to 4 fps, the number of parallel tube circuits can be determined.

The value of the film coefficient for water inside the tube can be found from Figs. 40 and 41.

The overall coefficient of heat transfer is expressed by the following:

$$U_o = \frac{1}{R_o + R_i + R_t + R_{fo} + R_{fi}} \text{ Btu/(h)(sq ft)(F°)}$$

where $R_o = \dfrac{1}{(h_o)(E)}$: h_o from Fig. 18 and 19
E = fin efficiency, Fig. 42

$R_i = \dfrac{A_o/A_i}{h_i}$: A_o = external finned surface/ft
A_i = inside tube surface/ft
h_i = from Figs. 40 and 41

$R_t = \dfrac{k}{t}$: k = tube wall thermal conductivity,
Btu/(h)(sq ft)(F°)/(ft)
t = tube wall thickness, ft

$R_{fo} = 0.002$: or other assumed value, depending on gas cleanliness
$R_{fi} = 0.001$: assumed for internal fouling factor

The total external (gas-side) heating surface required for a given set of conditions is $Q/(\text{Lm } \Delta T)(U_o)$,

where Q = Total required heat recovery, Btu/h
Lm ΔT = Log-mean temperature of the system
U_o = Overall heat transfer coefficient

After the heat transfer assembly has been determined by the method above, it must be checked for liquid-side pressure drop using Figs. 43, 44, and 45. If pressure drop is outside the design range, another trial must be made using a different tube size and more or fewer parallel tubes or both, as required. Usually not more than two trials need be made.

Multiple-tube circuits are always required for forced recirculation heat recovery boilers. It is important that water distribution be as equal as possible in all circuits. The high recirculation rate helps assure this goal. However, it is common to use restricting orifices or tubular restrictors at the inlet end of each tube circuit. Usually the tubular restrictor is more effective because a larger opening is possible than in an orifice plug or plate, which might easily be closed by dirt. A pressure drop of 5 psig at the circuit inlet does much to assure more equal water flow. The restrictor must be designed to be removable so that a clear opening to the tube is provided for tube cleaning.

Control of a heat recovery boiler may be accomplished by a gas bypass or by a surplus steam dump to a condenser. If a controllable gas bypass is used, it is common to have a controllable louver-type or butterfly valve at the boiler gas inlet and another in the bypass stack. They are controlled so that when one is open, the other is closed. Pneumatic or electric controls are used to activate the bypass system in response to steam pressure control.

When a steam dump valve is used, a pneumatically or electrically operated steam dump valve is located on the steam header and connected to the condenser. A steam pressure controller regulates the amount of steam diverted to the condenser.

Some gas turbine–steam turbine combined-cycle systems have no gas bypass control or steam dump. All the steam that can be generated by the heat recovery boiler is used in the steam turbine all the time so that the steam turbine carries all the load of which it is capable, under any given gas turbine load condition. This arrangement provides for optimum heat rate at all times. A gas bypass in this situation is either fully open or closed. It is fully closed when the steam system is used and fully open when it is not used. Sometimes large, guillotine-type shutoff gates are used at the boiler gas inlet to isolate more fully the heat recovery boiler from the gas turbine exhaust. Such gates usually require a motor drive arrangement to operate them open or closed.

Supplementary firing of heat recovery boilers is quite common, particularly in connection with combined-cycle power systems. Supplementary burners are located downstream of the inlet gate or valve. They come on when the gas and steam turbines are fully loaded and additional power is required. The burners receive their oxygen from the gas turbine exhaust gas, which is still sufficiently rich in oxygen to support combustion. The power realized from a system can be substantially increased by firing. The limit on supplementary firing is usually 1400°F. This allows conventional heat recovery boiler construction and does not require the use of water walls.

The gas turbine–steam turbine combined cycle will be used as an example for this section. A detailed analysis of heat recovery boiler design will not be made. Boiler design parameters will be established, however, for stated conditions. A straight, nonsupplementary-fired heat recovery boiler will be used, which offers the best overall heat rate. The following are gas turbine and steam conditions:

Gas turbine full continuous rating—10.4 MW
Specific fuel consumption (SFC)—0.71 lb/kWh
Simple-cycle heat rate—13,774 Btu/kW
Exhaust gas flow—113 pps
Exhaust gas temperature—992°F
Proposed steam temperature—800°F
Proposed steam pressure—600 psig
Feedwater temperature—225°F

Assume that all gas turbine exhaust goes to the heat recovery boiler at all times. The boiler will consist of a superheater, a high pressure evaporation section, an economizer, and a low pressure steam section for feedwater heating. Refer to Fig. 82 for the temperature-heat content diagram for the system. The pinchpoint on the diagram most affects boiler size and design. The pinchpoint temperature difference is the difference in temperature between the saturated steam temperature (489°F) and the gas temperature at that point. Good design places this differential between 20 and 50°F. In this case, 40°F will be used. Therefore, the unit will be designed to produce a gas temperature of 529°F at the pinchpoint.

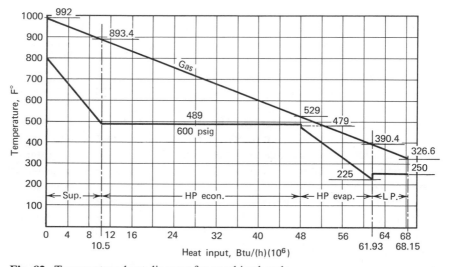

Fig. 82. Temperature-heat diagram for combined cycle.

Assuming no external losses, heat recovery from the exhaust temperature to the pinchpoint is

$(113)(3600)(0.255)(992 - 529) = 48,028,842$ Btu/h

Steam rate $= \dfrac{48,028,842}{1407 - 474.7} = 51,517$ lb/h

($1407 =$ enthalpy of steam at 600 psig, 800°F)

($474.7 =$ enthalpy of saturated water at 600 psig)

Superheater heat rate $= (51,517)(1407 - 1203)$
$= 10,509,468$ Btu/h

ΔT gas, superheater $10,509,468/(406,800)(0.262) = 98.6°F$

Gas temperature after superheater $= 893.4°F$

Evaporator heat rate, hp $=$ 48,028,842 $-$ 10,509,468
$$= 37,519,375 \text{ Btu/h}$$
Economizer heat rate $=$ (51,517)(463.4 $-$ 193)
$$= 13,930,197 \text{ Btu/h}$$
ΔT gas, economizer 13,930,197/(406,800)(.247) $=$ 138.6°F

Gas temperature after economizer $=$ 390.4°F

Feedwater heating and deaerating heat requirements are supplied by the low pressure section. The deaerator receives water from the condenser hot well at 115°F. Allowing 10% for the low pressure section itself, heat required is

(51,517 + 5152)(225 $-$ 115) $=$ 6,233,590 Btu/h
ΔT gas, L.P. section $=$ 6,233,590/(406,800)(.244) $=$ 62.8°F
Final stack temperature $=$ 390.4 $-$ 62.8 $=$ 327.6°F

Values of specific heat in the foregoing are taken from Fig. 2, Curve B. All terminal temperatures and heat rates for each of the four boiler sections must be determined. Heat recovery effectiveness $=$ (992 $-$ 327.6)/(992 $-$ 60) $=$ 71.29%.

The various points are summarized as follows:

	Superheater	Evaporator	Econ.	L.P. Section
Gas ΔT	98.6	364.4	138.6	62.8
Lm ΔT	285.5	157.7	96.6	117.9
Btu/h $\times$ 10^6	10.51	37.52	13.93	6.23

Tube sizes that can be used for this boiler are

Superheater—2-in. o.d.
hp Evaporator—2-in. o.d.
hp Economizer—$1\frac{1}{4}$- or $1\frac{1}{2}$-in. o.d.
L.P. section—2-in. o.d.

Tube walls are computed in accordance with requirements of the ASME Power Boiler Code, Section 1.

The data in the heat transfer and gas flow curves of Figs. 4, 18, 19, 40, 41, and 42 and the pressure drop curves of Figs. 4, 38, 39, and 43 are then used in conjunction with the terminal conditions established for the boiler (Fig. 82) to design the four boiler sections.

Therefore, 51,517 lb/h of steam at 600 psig and 800°F is available for power generation. Assume that condenser pressure is 3 in. Hg absolute maintained by steam jet ejectors. The theoretical water rate of a steam turbine under these conditions is 7.2 lb/kWh from Molier charts. The efficiency of a steam turbine generator and gear box can be taken at 72%. Therefore, the actual water rate is 7.2/.72 $=$ 10.0 lb/kWh. Power generation rate is 51,517/10 $=$ 5152 kW.

The total generating rate of both the gas and steam turbine generator is 15,552 kW.

The combined heat rate is now reduced from 13,774 Btu/(kW)(h) to 9211 Btu/(kW)(h), and overall cycle efficiency is 37%.

This example illustrates the benefits of recovered heat converted to additional power in a gas turbine cycle. Overall cycle efficiencies of up to 41% have been realized. Much depends on the mechanical efficiency of the gas turbine and also on the efficiency of the steam turbine generator. In the example cited, it is evident that as much heat has been recovered from the gas turbine exhaust as is practically feasible with a stack end approach temperature difference of 77.6°F, which is within the usual range of 20 to 100°F. Therefore, any additional gain must come from turbines of increased efficiency.

Assume that instead of a heat rate of 13,774 Btu/kWh, the gas turbine has a heat rate of 12,000 Btu/kWh and the steam turbine has an efficiency of 78% instead of 72%. Under these conditions, the combined heat rate is 7809 Btu/kWh, and overall cycle efficiency is 43.7%. Larger plants have a better chance for high efficiency because their larger turbines are more efficient.

COMBINED-CYCLE ECONOMICS

The economics of power plants varies considerably, depending on prevailing fuel costs, labor costs, and interest rates. In a combined-cycle plant, the addition of a steam cycle to the simple-cycle gas turbine will pay off, over a period of time, the cost of the entire plant. This is why fuel costs are so important. The saving of fuel from heat recovery is a prime objective in addition to increased power generation. The example cited in the previous section is also used in this analysis.

Gas turbine generator, 10.4 MW	$1,404,000
Gas turbine generator switchgear	75,000
Heat recovery boiler	244,705
Feedwater treatment equipment and pumps	20,000
Steam turbine generator, 5.15 MW	827,250
Steam turbogenerator switchgear	20,000
Steam condenser	75,000
Condenser auxiliaries	20,000
Miscellaneous items	50,000
Total Equipment	$2,735,955
Site preparation	100,000
Installation cost	410,000
Total Investment	$3,245,955

Total combined power—15.15 MW
Gas turbine fuel input—143,249,600 Btu/h
Combined SFC—9455 Btu/kWh

Power from steam turbine only—5.15 MW
Steam rate—51,500 lb/h
Equivalent fuel saved at boiler efficiency of 85%—72.893 × 10⁶ Btu/h
Fuel cost at $2.00/MBTU—$145.79/h
Annual fuel cost saving for plant at 70% load factor—$893,984

Owning Costs (based on target pay out of 8 years):
   Amortization, straight line $405,744
   Interest, average annual, 12%   219,102
                               $624,846

Operating Costs:
Electrical	$ 11,826
Maintenance, @ 1.0 mills/kWh	79,628
Insurance, 1%	32,460
Taxes, @ $1\frac{1}{2}$%	48,690
Water	2,000
Water treatment	5,000
Operators (3 shifts)	45,000
	$224,604
Total Owning and Operating Costs	$849,450

The value of fuel saved by the heat recovery and steam turbine supplementary-power system was $893,984, which compares favorably with projected annual owning and operating costs. Therefore, the entire system can be written off within an 8-year period by fuel savings alone. This frees a substantial portion of revenues for profit, dividends, and financing of future expansion.

Plants of larger capacity can be written off in a shorter time, some as short as five years, depending on plant characteristics. The SFC (Specific Fuel Consumption) rate of 9455 Btu/kWh should be especially noted. Heat rates for straight-steam plants in the 10 to 15 MW size range from 12,000 to 14,000 Btu/kWh and diesel engine plants about 10,000 to 10,500 Btu/kWh. The plant cited in this example has a heat rate lower than steam plants of a much larger size. For example, in the 80-MW size, a steam plant will have a heat rate of about 10,200 Btu/kWh, while a combined-cycle plant can be as low as 8800 Btu/kWh. In summary, within the 10-MW to 80-MW range, combined-cycle designs are less expensive to buy and install than diesel or steam turbine plants and, depending on size, have anywhere from 10 to 50% better heat rate.

ORGANIC RANKINE CYCLE COMBINED POWER PLANT

While the organic working fluid power plant is not, strictly speaking, in the steam category, it is a vapor cycle and therefore is treated briefly in this chapter. Although organic Rankine cycle combined plant is not in general use as of this writing, it has promise for future applications.

A Rankine cycle power generation system with organic heat transfer working fluid can economically utilize low temperature ($\leqslant 1000°F$) heat sources for the generation of electric power. For example, a regenerated gas turbine of 36.6% efficiency can be improved to 47% efficiency by the addition of an organic Rankine cycle bottoming plant operated from gas turbine exhaust heat.

The use of this system as a bottoming plant to the gas turbine is potentially the optimum method of attaining these high efficiency goals while still providing low-cost electric power. For every case studied, given the basic gas turbine unit, the organic bottoming cycle resulted in substantially higher efficiency than the equivalent steam bottoming cycle. The best combination turns out to be a regenerated gas turbine coupled to an organic bottoming cycle.

Several different working fluids investigated since 1963 might be suitable for power generation application. The increasing availability of these new fluids has spurred the design of the modern organic Rankine cycle, whose present state of technology permits realistic development of organic systems.

In the organic Rankine cycle, superheated vapor is expanded through either a reciprocating engine or a turbine. The low pressure vapor containing substantial superheat passes through the gas side of a regenerator, transferring heat to the boiler feed liquid. The vapor from the regenerator is condensed and pumped to boiler pressure, preheated in the regenerator, and then passed through the boiler. Turbine expanders are preferred in cases where constant speed can be maintained.

Turbine expansion with a suitable fluid occurs solely in the superheated vapor region, whereas in the simple steam cycle, expansion invariably results in wet steam in the final stages. The presence of moisture can result in blade erosion and loss of efficiency. The organic fluid's characteristics are such that the turbine can be a single-stage, axial-flow, impulse type, in contrast to steam turbines that require multiple stages for maximum turbine efficiency.

The many factors to be considered in choosing a suitable working fluid include: thermal stability, compatibility, safety, cost, physical characteristics, and thermodynamic characteristics.

One working fluid with a unique combination of desirable characteristics is a mixture of trifluorethanol (CF_3CH_2OH)—85 mole %, and water, 15 mole %. The characteristics of this fluid are as follows:

Chemical composition CF_3CH_2OH—85 mole %
Water—15 mole %
Average molecular weight—87.74
Freezing point—82°F
Boiling point—165°F
Flammability—none
Liquid density—1.25 gm/cm^3
Lubricant—refrigeration oil
Thermal stability—550°F (potentially to 650°F)

The temperature entropy diagram is illustrated by Fig. 83. This fluid seems to be ideally suited to Rankine cycle systems having a wide power range, and for expander inlet temperatures from 300 to 650°F. The fluid is compatible with all common construction materials, requiring no chemical control whatsoever for corrosion inhibition.

Use of an organic working fluid provides the maximum amount of recovered energy practically obtainable from a low temperature gas source. This effect is a result of the low heat of vaporization of the organic working fluid relative to water, which reduces the temperature differential in the boiler and permits the organic cycle to approach close to thermodynamically reversible operation. Researchers have found that 25 to 50% more work, depending on gas temperature, can be obtained from the gas stream by use of an organic working fluid. Cycle characteristics for unrecuperated and recuperated cycles are as follows:

Unrecuperated gas turbine/organic cycle gas turbine
    Model number   GE G5211
    Power (80°F, 1000 ft)   15,250 kW
    SFC (Specific Fuel Consumption)   0.77 lb/kWh
    Heat Rate (hhv)   15,000 Btu/kWh
    η_{OA} (hhv)   22.8%
    Exhaust gas flow rate 7.09×10^5 lb/h
    Exhaust gas temperature   930°F
Organic Rankine cycle bottoming plant
    Exhaust gas temperature from boiler   250°F
    Organic fluid flow rate 4.33×10^5 lb/h
    Turbine
        Type   single stage, axial
        rpm   3600
        Tip diameter   5.5 ft
        Blade height   5 in.
        η_{TH} 0.85
        η_M 0.95

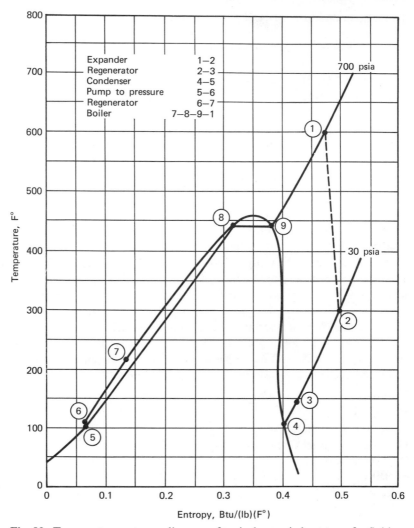

Fig. 83. Temperature entropy diagram of typical organic heat transfer fluid.

η_{OA}	0.808
Power	10,050 kW
Overall plant characteristics	
Gas turbine power	15,250 kW
Organic turbine power	10,050 kW
Total power	25,300 kW

Percentage increase in power	66%
η_{OA} binary plant	37.8%
Heat rate, binary plant	9050 Btu/kW
Recuperated gas turbine/organic cycle gas turbine	
Model number	GE PG7821
Power (59°F, sea level)	66,600 kW
SFC (Specific Fuel Consumption)	0.593 lb/kWh
Heat rate (hhv)	11,580 Btu/kWh
η_{OA} (hhv)	29.5%
Exhaust flow	1.896×10^6 lb/h
Exhaust gas temperature	1040°F
Organic Rankine cycle bottoming plant	
Exhaust gas temperature from boiler	250°F
Organic fluid flow	1.345×10^6 lb/h
Turbine	
Type	single stage, axial
rpm	3600
Tip diameter	7.4 ft
Blade height	9 in.
η_{TH}	.85
η_M	.95
η_{OA}	.808
Power	31,200 kW
Overall plant characteristics	
Gas turbine power	66,600 kW
Organic turbine power	31,200 kW
Total power	97,800 kW
Percentage increase in power	47%
η_{OH} binary plant	43.3
Heat rate, binary plant	7820 Btu/kWh

The preceding data make clear the desirability of pursuing the organic fluid binary power cycle for future applications. This cycle should also be considered for the smaller power ranges, from 5000 to 10,000 kW base load, and for closed-cycle systems for automated operation in remote regions, such as gas compressor stations and offshore applications. The binary cycle is a good example of heat recovery for power generation.

CONCENTRIC PIPE HEAT EXCHANGER

CONFIGURATION

The concentric pipe heat exchanger is basically composed of two pipes or tubes, one within the other, with fluids flowing within the annulus and the inner pipe in counterflow relation. This form of heat exchanger is actually a perfect counterflow device. One of its problems, however, is that fluid flow is parallel to the pipe wall, which results in lower coefficients of heat transfer than in crossflow designs, thereby requiring more surface area.

Cleaning

Another problem is that the annulus between the inner pipe and the outer pipe is generally inaccessible for cleaning unless the entire inner pipe is removable by means of a flanged construction at the ends. Concentric or double-pipe exchangers are seldom made of small-diameter tubes, although these are available for specialized applications. Usually, they are made of larger pipe sizes ($1\frac{1}{2}$ in. and more) and in lengths of up to 40 ft. The length raises the problem of differential expansion between the inner pipe and the outer pipe, which will nearly always be at different temperatures.

Differential Expansion

Differential expansion can be handled either by designing the exchanger using return bends at one end of each pair of pipes or by using packing boxes or bellows at one end. The return bend design is the more common because it allows the two inner pipes of a pair to be rigidly attached at one end to expand and contract at the return bend end within an enlarged retaining box deep enough to permit movement. Refer to Fig. 84 for a typical construction for the ends.

Construction materials for double-pipe exchangers vary with application. Different materials may be used for the inner and outer pipes, depending on

190

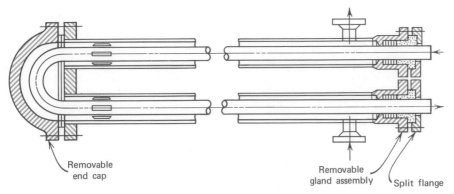

Fig. 84. Typical double-pipe heat exchanger and construction.

service and fluid characteristics. Material may be selected to meet either chemical or mechanical conditions, such as abrasion caused by sludge content. For example, sewage sludge may contain ground glass, steel particles, sand, paper, and other materials that cause wear. Abrasion or erosion is sometimes so severe that schedule 80 pipe is used in applications that would otherwise use only schedule 40 pipe. Where there is a choice, the abrasive material flows through the inner pipe, which is more accessible for service and cleaning.

APPLICATION

The concentric pipe heat exchanger is often used as a heat recovery device in process applications, such as heat treatment of sewage sludge and other liquid-to-liquid heat recovery systems. The sewage sludge application is an excellent example of liquid-to-liquid heat recovery. In this system, the heat exchanger is used to transfer the heat content of the hot sludge coming from the reactor to the cold sludge coming in to the reactor. Terminal temperature differences in this application may be as low as 60°F. Naturally, the smaller the terminal temperature differences, the more heating surface is needed exponentially. The temperature conditions in a typical sewage sludge heat treatment system are as follows:

Cold sludge in—60°F
Hot sludge out—320°F
Hot sludge in—380°F
Cold sludge out—120°F

The conditions listed above are true if flows are equal on both sides. In the sewage sludge application, the heat recovery process may be either a direct system (that is, sludge-to-sludge) or a sludge-to-water and water-to-sludge system. The latter allows operation at lower terminal temperature differences and hence helps to prevent formation of deposits on the tube wall caused by nucleate boiling. Figure 85 schematically illustrates a water loop system. The simple single-element double-pipe exchanger is, of course, simpler and less expensive than the water loop system. However, its disadvantage is that it has sludge in both the central pipe and the annulus, and the annulus is very difficult to clean. Only an acid wash can be used, which is not always fully effective. In the water loop system, water flows through the annulus and sludge through the inner pipe. For inspection and cleaning, the inner pipe can be made accessible from both ends by removable flanges or removable end caps.

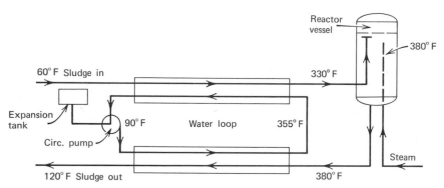

Fig. 85. Typical schematic, double-pipe heat exchanger water loop regenerative system.

SCRAPED SURFACE HEAT EXCHANGER

In another form, the concentric pipe heat exchanger is made with a continuously scraped inner pipe. This form is used when viscous material is pumped through the inner pipe. Some applications for the scraped surface exchanger are listed below:

1. Heat transfer with crystallization
2. Heat transfer with severe fouling
3. Heat transfer with solvent extraction
4. Heat transfer with continuous mixing and conveying
5. Heat transfer with high viscosity

Figure 86 illustrates a typical scraped surface heat exchanger. This type of exchanger gives continuously uniform heat transfer rates with process fluids that would quickly foul up conventional heat exchangers. These exchangers permit process industries to heat or cool crystallizing or viscous fluids without loss in heat transfer rates.

Fig. 86. Scraped surface heat exchanger. Courtesy Henry Vogt Machine Co.

In a scraped surface exchanger, the process fluid flows through a series of pipes connected by crossovers and arranged on common supports. The inner surfaces of the inner pipes are continuously scraped clean by rotating flights of scraper blades. The outer surfaces of the inner pipe (the annulus) is heated or cooled by a heat transfer medium that does not contaminate or foul appreciably.

The heart of the scraped surface exchanger is the assembly of the scraper elements, which continuously clean the inner pipe surfaces and agitate the process fluid. The scraper elements consist of a series of blades, each attached by several springs to a central rotating shaft. The springs hold the scraper blades firmly but flexibly against the inner pipe wall. A series of short overlapping blades permits scraper flight to conform to irregularities in the pipe wall. The drive end of the scraper flight engages the drive shaft, usually through a slip coupling. Sprockets on the scraper shafts are driven by a continuous roller chain that engages all the sprockets and the drive motor.

Scraped surface exchangers are the usual answer to heat transfer problems associated with crystallization, high viscosity, and thermal reactions. As applied to crystallization, the scraper blades continuously remove crystal formations from the cooling surface. Scrapers maintain the crystal slurry in suspension to promote its removal from the exchanger. Exchangers with a small annulus, a relatively high shaft speed, and a small volume or hold up per unit of heat transfer surface result in highest heat transfer efficiency but generally yield a small, uniform, crystal size. This approach is sometimes desirable on material requiring supercooling before discharge from the exchanger. For crystals of larger particle size, a lower shaft speed is used, and hold up time in the exchanger is increased, tending to produce larger crystals.

Problems involving high viscosity are also easily resolved. In conventional heat exchangers, the laminar flow viscous fluids resist heat transfer. The scrapers break up the laminar flow and mix the wall film into the main stream. Viscous polymers, wax slurries, and equivalent fluids can be cooled in scraped units at substantial transfer rates with comparatively low pressure drop. Materials can be treated with viscosities as high as 100,000 Cp. Higher viscosities can be treated with special exchanger configurations.

Continuous agitation from scraper flights adapt these exchangers well to solvent extraction or slurry reactions associated with heat transfer requirements.

FOULING

The formation of scale deposits in concentric pipe exchangers is a function of the heat transfer rate across the tube wall, fluid velocity, fluid shear on the particular material deposited, chemical makeup of the scale forming ingredients, film temperature, and film pressure. It is possible to determine the fluid velocity required to eventually cause the erosion effects of fluid shear to balance the rate of deposit, so that a tube will scale up to a point, with little or no additional deposit thereafter. Other important factors are temperature and pressure of the fluid film in contact with the tube wall. System pressure must always be maintained well above the boiling pressure for the highest film temperature. This will largely reduce or eliminate deposits caused by local nucleate boiling. Any drop in pressure caused by cavitation, slugging, restrictions, and so on may cause local nucleate boiling and result in deposits of solids. In many closed systems where boiling at high temperature is to be avoided, the system can be pressurized by a surge tank under nitrogen gas pressure supplied from bottles and regulated to the required pressure.

Extended surfaces are used on the outer diameter of the inner pipe only when the fluid in the annulus has a much lower coefficient of heat transfer than the fluid in the inner pipe. In this case, the extended surface consists of longitudinal fins. For example, if the external coefficient is only one-fifth of the internal coefficient, then external effective extended surface can be at least five times the internal surface. This is the case with oil heaters.

Fluid velocity on both sides has an important relation to the overall heat transfer coefficient. Velocity on both sides must be high enough to promote good heat transfer yet not produce excessive pressure drop. When abrasive sludges such as sewage sludge are handled, velocities much in excess of 4 fps produce increased erosion. The base value of the heat transfer coefficient varies approximately as $V^{0.8}$. This relation indicates that the coefficient increases indefinitely in an exponential relation. There are, however, three other factors of resistance that moderate this relation: (1) tube wall resistance, (2) internal fouling, and (3) external fouling. Tube wall resistance is a known factor related to the thermal conductivity of the tube wall and tube thickness. Fouling factors, however, are more indefinite because of variations in fluid composition, scaling rates, oil and grease content, and so on. Increases in velocity can reduce the extent of fouling due to fluid shear. Conceivably, through increases in fluid velocity and the accompanying increases in fluid shear, the fouling asymptotic resistance could be so manipulated as to add only a tolerable 5 to 10% oversurface to the required clean surface. Then, the exchanger could be placed on stream and allowed to foul to a tolerable antic-ipated asymptote. Since it could not foul any further, it would never require cleaning. Liquid shear has not been examined as a cleaning agent, but merely as a suppressant of deposition at the very moments that determine whether a particle attaches to the heat transfer surface or is swept downstream by the liquid.

With abrasive fluids that limit velocity, the concept of fluid shear cannot operate, and some realistic values of fouling factor must be taken. There is no hard rule for determining fouling factors since they are usually determined by experience. Table 6 in Chapter 2 gives practical fouling factors for various conditions. The application of fouling factors tends to moderate the overall heat transfer coefficient, as illustrated by Fig. 87, which is based on a con-stantly increasing equal fluid velocity on both sides of a concentric pipe heat exchanger, at fixed fouling factors for each side.

A good example for calculating a concentric pipe heat exchanger is a typical sewage sludge application. The exchanger is operated in a sludge-to-sludge mode with cold sludge in the annulus and hot sludge in the inner pipe. The application is typical of liquid-to-liquid heat recovery. Heat transfer

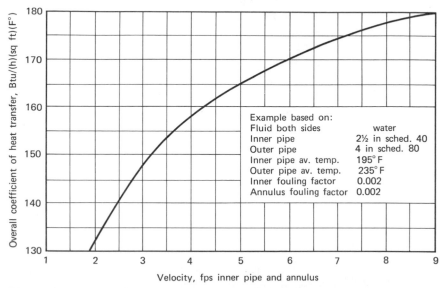

Fig. 87. Effect of velocity on overall heat transfer coefficient.

coefficients for both sides are taken as water, because sewage sludge is 93 to 95% water, with the balance being solids and organic materials. Figure 88 illustrates a typical schematic and heat balance.

Operating conditions:
Sludge flow to heat exchanger—3500 gph
Sludge temperature to heat exchanger—60°F
Sludge temperature after heat exchanger—330°F
Reactor temperature—380°F

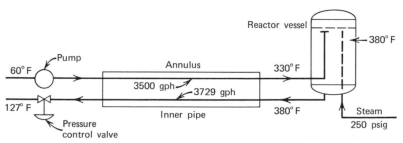

Fig. 88. Typical schematic, double-pipe sludge-to-sludge heat recovery system.

Sludge out of reactor—3729 gph
Sludge temperature to heat exchanger—380°F
Sludge temperature out of heat exchanger—127°F
Heat added in reactor $= (3500)(8.4)(380 - 330)$
$$= 1,470,000 \text{ Btu/h}$$
Allow for 10% heat loss in the system.
Steam to reactor $= 1,470,000/(.90)(1201.7 - 354) = 1927 \text{ lb/h}$
$$= 229 \text{ gph}$$
Steam added to the reactor is not recovered as condensate and is added to the sludge. Flow out of the reactor $= 3500 + 229 = 3729$ gph.
Heat exchange rate $= (3500)(8.4)(330 - 60)$
$$= 7,938,000 \text{ Btu/h}$$
Hot side $\Delta T = 7,938,000/(3729)(8.4) = 253°F$
Hot side outlet temperature $= 380 - 253 = 127°F$
System Lm $\Delta T = 58°F$
Assume inner pipe velocity of approximately 4.25 fps, $2\frac{1}{2}$ in.
Schedule 40 pipe (0.203 wall) will produce a velocity of 4.16 fps at 3729 gph.
The required annular area for the same velocity in the annulus $= 4.788$ sq in.
Approximate outer pipe i.d. $= 3.683$ in. Select 4-in. Schedule 80 (.337 wall) outer pipe, i.d. $= 3.826$ in. Annulus velocity $= 3.74$ fps.

Heat Transfer, inner pipe:
From Fig. 40, for velocity of 4.16 fps
$h = 990 \text{ Btu/(h)(sq ft)(F}°)$
From Fig. 41, for i.d. $= 2.469$ in., $F_d = 0.83$
From Fig. 41, for average temperature 253°F, $F_t = 1.60$
$h_i = (990)(0.83)(1.60) = 1315 \text{ Btu/(h)(sq ft)(F}°)$

Heat transfer, annulus:
From Fig. 40, for velocity 3.74 fps
$h = 900 \text{ Btu/(h)(sq ft)(F}°)$
From Fig. 41, for average temperature 195°F, $F_t = 1.20$
Equivalent diameter $= \dfrac{3.826^2 - 2.875^2}{2.875} = 2.22$
From Fig. 41, for $d_e = 2.22$, $F_d = 0.85$
$h_o = (900)(1.20)(0.85) = 918 \text{ Btu/(h)(sq ft)(F}°)$
h_w (tube wall) $= 25/(.203/12) = 1478 \text{ Btu/(h)(sq ft)(F}°)$

$$R_i = \frac{1}{1315} = 0.0007604$$

$$R_o = \frac{1}{918} = 0.0010893$$

$$R_w = \frac{1}{1478} = 0.0006766$$

R_{fi} (internal fouling) $= 0.002$
R_{fo} (external fouling) $= 0.002$
$R_t = 0.0065263$
$$U_o = \frac{1}{R_t} = 153 \text{ Btu/(h)(sq ft)(F°)}$$
Surface $= 7,938,000/(153)(58) = 895$ sq ft

Based on the inside surface of the inner pipe of 0.6464 sq ft/ft, require 895/0.6464 = 1385 ft of effective double pipe.

Either 35-to-40-ft lengths or 46-to-30-ft lengths connected in series satisfy the requirements. The individual elements can be mounted in a rack or frame to form one unit. The pipes may be either individually insulated or the entire assembly may be enclosed in a prefabricated, insulated, panel casing. The elements may be mounted either vertically or horizontally. The vertical arrangement saves floor space; however, the horizontal arrangement makes the end connections more accessible.

INSULATION

Heat Loss

Proper insulation of heat recovery equipment is important for efficient recovery of heat from any source, not only from an economic standpoint but also for personnel comfort and safety.

Heat Loss from Furnace Walls

The calculation of heat transfer through insulation follows the principles outlined for conduction through a composite wall. In a boiler furnace with tube-to-tube walls, the hot face temperature of the insulation may be taken as saturation temperature of the water in the tubes. If the inner face of the furnace wall is refractory, with or without cooling by spaced tubes, the hot face temperature of the insulation must be calculated or estimated from a knowledge of radiation and convection heat transfer on the gas side of the furnace wall, or from empirical data. Fortunately, a considerable error in calculating the hot face temperature of insulation involves a much smaller error in cold face temperature, and it may not introduce significant error into heat loss calculation. However, an error in hot face temperature calculation may lead to inappropriate selection of materials.

Heat loss to the surroundings and cold surface temperature decrease as the thickness of the insulation increases. When the insulation is thick, the change in cold surface temperature for an increase in thickness is small, whereas the cost of the insulation increases steadily with thickness. Standard commercial thicknesses of insulating materials should be used in the composite wall.

Low ambient air temperature and high air velocities tend to reduce the cold face temperature, although they have only a small effect on total heat loss, since surface film resistance is a minor part of total insulation resistance to heat flow. Combined heat loss rates (radiation plus convection) expressed

199

in Btu/(sq ft)(h) are given in Fig. 89 for various temperature differences and air velocities. The effect of surface film resistance on casing temperature and heat loss through casings is shown in Fig. 90.

Refractory or insulating materials suitable for high temperature applications is usually more expensive and less effective as insulation than low temperature materials. It is, therefore, customary and economical to use

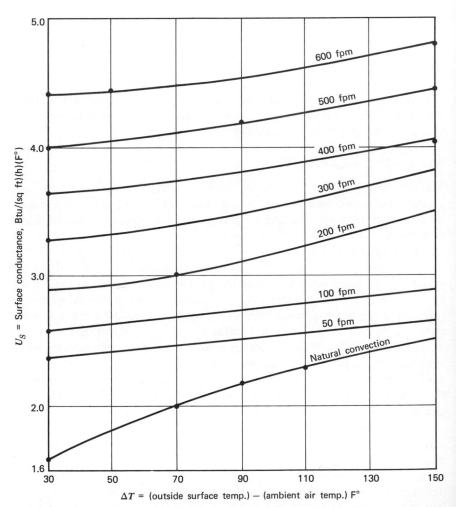

Fig. 89. Heat loss factor U_s for furnace walls, natural convection and at varied air velocities.

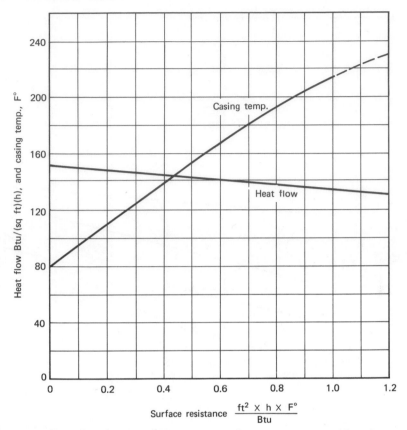

Fig. 90. Effect of surface film resistance on surface temperature and heat loss rate.

several layers of insulation—the lowest cost and most effective insulation in cool zones and the higher cost materials only where temperature exceeds permissible operating limits for low temperature materials. Thermal conductivities for refractory and insulating materials at suitable temperatures, are given in Fig. 91.

To maintain satisfactory working conditions around an indoor boiler, the insulation must be thick enough to keep outside skin temperatures reasonably low and to prevent an excessive increase in boiler room temperature. A cold face temperature of 130 to 160°F is usually considered satisfactory for an indoor installation. Heat losses corresponding to these temperatures range between 70 and 180 Btu/(sq ft)(h), which can be readily absorbed by the air circulation generally provided in modern boiler rooms.

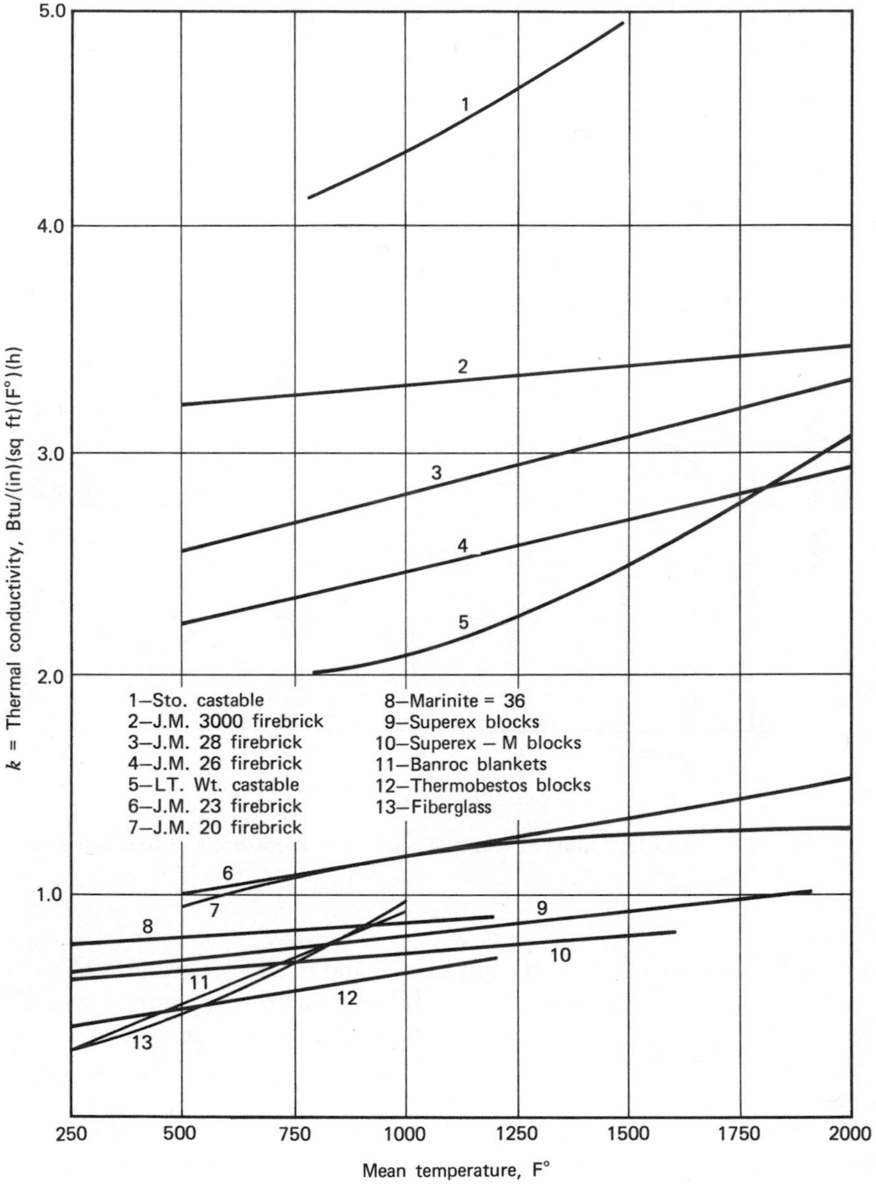

5.0
4.0
3.0
2.0
1.0
k = Thermal conductivity, Btu/(in)(sq ft)(F°)(h)
1
2
3
4
5
6
7
8
9
10
11
12
13
1–Sto. castable
2–J.M. 3000 firebrick
3–J.M. 28 firebrick
4–J.M. 26 firebrick
5–LT. Wt. castable
6–J.M. 23 firebrick
7–J.M. 20 firebrick
8–Marinite = 36
9–Superex blocks
10–Superex – M blocks
11–Banroc blankets
12–Thermobestos blocks
13–Fiberglass
250
500
750
1000
1250
1500
1750
2000
Mean temperature, F°

Fig. 91. Thermal conductivities and temperatures for refractory and insulating materials.

Insulating a boiler to reduce heat loss to an amount readily absorbable by the total volume of room air does not in itself assure comfortable working conditions. Good air circulation around all parts of the boiler is also necessary to prevent accumulation of heat in areas frequented by operating personnel. This can be aided by substitution of grating for solid floors, by ample aisle space between boilers, by the location of fans to assist air circulation around boilers, and by the addition of ventilating equipment to assure adequate air change.

Fortunately, on modern units good ventilation does not greatly increase overall heat loss. Air velocity affects surface conductance ($q/s \div$ temperature difference); this can be verified by data from Fig. 89. However, surface conductance is only a small part of the total resistance to air flow. For example, an increase in air velocity from 1 to 10 fps, for the conditions given in Fig. 92, increases heat loss rate through the wall by only 3 Btu, from 144 to 147 Btu/(sq ft)(h). This is about 2%, for a tenfold increase in air velocity.

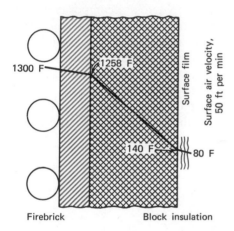

Fig. 92. Temperature gradients through tube and brick wall.

Effect of Surface Film Resistance on Surface Temperature and Heat Loss

Unlike heat loss, outer surface temperature is affected considerably by surrounding air conditions. From Fig. 90 it is seen that a considerable change in surface film resistance will cause an appreciable change in lagging or surface temperature while not affecting to any extent heat loss through the wall. Dead cavities should be avoided, particularly where operators must go, and ventilating ducts should be installed in such areas. Increased insulation thickness will not significantly reduce surface temperature in such spaces.

Insulating material should be selected to be adaptable. It must (1) be easily applied; (2) have heat resistant qualities sufficient to withstand successfully the highest temperatures to which it will be subjected; (3) be sufficiently strong and durable to assure long life. Adaptability also depends on many other considerations incidental to the particular application. Thermal conductivities and temperatures for refractory and insulating material is shown on Fig. 91.

The insulating value of a commercial insulation depends primarily on the small voids it contains. To be most effective, voids must be enclosed and so small that circulation within them and radiation across them will be minimal. Small void size is particularly necessary at high temperatures because of the rapid increase in convection and radiation with temperature rise. The usual materials are asbestos, magnesium carbonate, diatomaceous silica, mineral wool, refractory clays, and fiberglass.

Composite Walls

Calculation of heat flow and temperature drop across a composite wall of refractory, steel and insulation is at best a trial and error procedure that combines thermal resistances of different materials (heat flow paths in series) into one overall resistance. Refer to Fig. 93, representing a composite wall of three different materials. Assume that Fig. 93 is the section through a flue that is refractory lined inside and insulated on the outside.

In addition to the thermal resistance (reciprocal of conductance) of refractory, steel, and insulation, heat flow through the flue must also be transferred

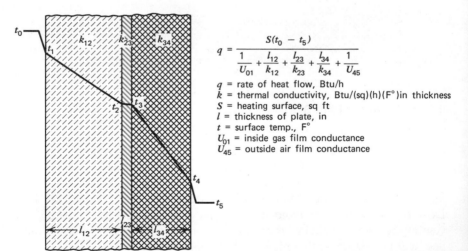

$$q = \frac{S(t_0 - t_5)}{\dfrac{1}{U_{01}} + \dfrac{l_{12}}{k_{12}} + \dfrac{l_{23}}{k_{23}} + \dfrac{l_{34}}{k_{34}} + \dfrac{1}{U_{45}}}$$

q = rate of heat flow, Btu/h
k = thermal conductivity, Btu/(sq)(h)(F°)in thickness
S = heating surface, sq ft
l = thickness of plate, in
t = surface temp., F°
U_{01} = inside gas film conductance
U_{45} = outside air film conductance

Fig. 93. Diagram of composite wall.

through gas films on both sides of the composite wall. Evaluation of gas film conductances is discussed in detail under the section convection heat transfer, but it should be recognized that the thermal resistance of the gas films must be combined with the thermal resistance of solid materials. Large-diameter ducts and flues or casings can be considered vertical plates for practical purposes. Gas or air flow inside and outside will probably be in natural convection. Figures 22 and 23 can be used for estimating values of thermal resistance of the inside and outside gas films. Use the following equation to determine overall thermal resistance:

$$U_o = \frac{1}{\dfrac{1}{U_1} + \dfrac{1_1}{k_1} + \dfrac{1_2}{k_2} + \dfrac{1_3}{k_3} + \dfrac{1}{U_2}} \tag{40}$$

This equation can be used for any number of wall components by simply extending the denominator to suit.

U_1 = hot-side gas film thermal resistance
U_2 = cold-side gas film thermal resistance
U_o = overall system thermal resistance
1_1 = wall thickness, first component, in.
1_2 = wall thickness, second component, in.
1_3 = wall thickness, third component, in.
k_1 = thermal conductivity, Btu/(sq ft)(h)(F°)(in.), first component wall
k_2 = thermal conductivity, Btu/(sq ft)(h)(F°)(in.), second component wall
k_3 = thermal conductivity, Btu/(sq ft)(h)(F°)(in.), third component wall

To determine whether correct thermal conductivity was assumed and temperature level was within allowable operating limits of the material, it is necessary to calculate temperatures on both sides of each material. This involves a step-by-step solution of the following equation:

$$q = U_0 S \, \Delta T \tag{41}$$

where q = rate of heat flow, Btu/h
U_0 = overall, or combined, conductance, Btu/(sq ft)(h)(F°)
S = surface involved in heat transfer, sq ft
Δt = temperature difference causing heat flow, F°
R = $1/U_o$ = combined resistance of heat flow path, sq ft/(h)(F°)(Btu)

$\Delta t = \dfrac{q}{US}$ = Referring to Fig. 93, the conditions for each component are as follows:

$$t_0 - t_1 = \frac{q}{U_1 S} \tag{41a}$$

$$t_1 - t_2 = \frac{q}{\dfrac{k_1}{1_1} S} \tag{41b}$$

$$t_2 - t_3 = \frac{q}{\dfrac{k_2}{1_2} S} \tag{41c}$$

$$t_3 - t_4 = \frac{q}{\dfrac{k_3}{1_3} S} \tag{41d}$$

$$t_4 - t_5 = \frac{q}{U_2 S} \tag{41e}$$

The preceding method must be used on a trial-and-error basis to find the optimum combination of materials. However, experience with a variety of materials reduces the number of trials to a minimum.

For any case in which heat flows through successive flat layers of material, the reciprocal of the overall conductance equals the sum of the individual resistance. If the successive layers of material do not make good thermal contact with each other, result will be interface resistance because of the air space or film. These resistances may be neglected in composite walls of insulating materials, but they become relevant considerations and must be included in calculations if the resistances of the layers are small in comparison with the interface resistances, such as in heat transfer through a boiler tube with an oxide deposit on the inside.

When heat is conducted radially through a cylindrical wall, as in heat flow through the wall of a steam line from the inside to the outside of a pipe, the flat plate must be modified, since the heat flow surface area S is no longer a constant, but increases as the distance from the center of the pipe increases. If the outside surface area of the pipe is used for the value of S, then the thickness of the pipe wall, corresponding to "1" in the flat plate equation, must be replaced by the equivalent thickness 1_e, given by Equation 42:

$$1_e = 0.50 \, d_o \, \log_e \left(\frac{d_o}{d_i} \right) \tag{42}$$

where 1_e = equivalent thickness, in.
$\quad\quad d_o$ = outside pipe diameter, in.
$\quad\quad d_i$ = inside pipe diameter, in.

Therefore, for successive cylindrical walls, substitute 1_e for 1 in Equation 41 and in the equations for each individual component.

Insulation requirements for standard pipe sizes are usually set forth by insulation manufacturers. Table 16 shows insulation requirements for steam pipes insulated with hydrous calcium silicate bonded with asbestos fibers. This completely inorganic material is good for pipe temperatures to 1200°F.

TABLE 16 Thermobestos, Insulation Thickness for Pipe (F°)

Nominal Pipe Size, in.	100 to 199	200 to 299	300 to 399	400 to 499	500 to 599	600 to 699	700 to 799	800 to 899	900 to 999	1000 to 1099	1100 to 1200
$1\frac{1}{2}$ or less	1	1	$1\frac{1}{2}$	2	2	$2\frac{1}{2}$	$2\frac{1}{2}$	$2\frac{1}{2}$	3	3	3
2	1	1	$1\frac{1}{2}$	2	2	$2\frac{1}{2}$	$2\frac{1}{2}$	3	3	$3\frac{1}{2}$	$3\frac{1}{2}$
$2\frac{1}{2}$	1	1	$1\frac{1}{2}$	2	2	$2\frac{1}{2}$	$2\frac{1}{2}$	3	3	$3\frac{1}{2}$	$3\frac{1}{2}$
3	1	1	$1\frac{1}{2}$	2	2	$2\frac{1}{2}$	3	3	3	$3\frac{1}{2}$	$3\frac{1}{2}$
$3\frac{1}{2}$	1	1	$1\frac{1}{2}$	2	$2\frac{1}{2}$	$2\frac{1}{2}$	3	3	$3\frac{1}{2}$	$3\frac{1}{2}$	$3\frac{1}{2}$
4	1	1	$1\frac{1}{2}$	2	$2\frac{1}{2}$	$2\frac{1}{2}$	3	3	$3\frac{1}{2}$	$3\frac{1}{2}$	4
$4\frac{1}{2}$	1	1	$1\frac{1}{2}$	2	$2\frac{1}{2}$	$2\frac{1}{2}$	3	3	$3\frac{1}{2}$	$3\frac{1}{2}$	4
5	1	$1\frac{1}{2}$	$1\frac{1}{2}$	2	$2\frac{1}{2}$	$2\frac{1}{2}$	3	$3\frac{1}{2}$	$3\frac{1}{2}$	4	4
6	1	$1\frac{1}{2}$	2	2	$2\frac{1}{2}$	3	3	$3\frac{1}{2}$	$3\frac{1}{2}$	4	4
7	$1\frac{1}{2}$	$1\frac{1}{2}$	2	2	$2\frac{1}{2}$	3	3	$3\frac{1}{2}$	$3\frac{1}{2}$	4	4
8	$1\frac{1}{2}$	$1\frac{1}{2}$	2	2	$2\frac{1}{2}$	3	$3\frac{1}{2}$	$3\frac{1}{2}$	4	4	$4\frac{1}{2}$
9	$1\frac{1}{2}$	$1\frac{1}{2}$	2	$2\frac{1}{2}$	$2\frac{1}{2}$	3	$3\frac{1}{2}$	$3\frac{1}{2}$	4	4	$4\frac{1}{2}$
10	$1\frac{1}{2}$	$1\frac{1}{2}$	2	$2\frac{1}{2}$	$2\frac{1}{2}$	3	$3\frac{1}{2}$	$3\frac{1}{2}$	4	4	$4\frac{1}{2}$
12	$1\frac{1}{2}$	$1\frac{1}{2}$	2	$2\frac{1}{2}$	3	3	$3\frac{1}{2}$	4	4	$4\frac{1}{2}$	$4\frac{1}{2}$
14 or more	$1\frac{1}{2}$	$1\frac{1}{2}$	2	$2\frac{1}{2}$	3	3	$3\frac{1}{2}$	4	4	$4\frac{1}{2}$	$4\frac{1}{2}$

A less expensive inorganic pipe insulating material made of hydrated basic carbonate of magnesia bonded with asbestos fibers, known as 85% Magnesia, is used for metal temperatures to 600°F. This material is also chemically stable and durable. Thermal conductivity ranges from 0.35 Btu/(in.)(sq ft) (F°)(h) at 100°F to 0.46 at 400°F. Table 17 gives recommended thicknesses of 85% Magnesia insulating material for various pipe sizes and pipe temperatures for steam systems.

Refractory materials are nearly always used for the inside lining of large heat recovery boilers. Either castable, brick, or block materials are used. Special precautions must be taken to assure air-tightness of a setting. Modern heat recovery boilers have an airtight steel casing on which the refractory is laid, poured, or gunned. Refractory walls susceptible to leaks caused by hairline cracking or improper mortaring between joints can cause heat loss. A pressure head on the hot side of the refractory, which occurs quite often in furnace roofs, causes a flow of hot gas across the wall. Consequently, gases transport their heat content by convection through the wall. This increases

TABLE 17 85% Magnesia Insulation Thickness for Pipe (F°)

Nominal Pipe Size, in.	100 to 199	200 to 299	300 to 399	400 to 499	500 to 600
$1\frac{1}{2}$ or less	1	1	$1\frac{1}{2}$	2	2
2	1	1	$1\frac{1}{2}$	2	2
$2\frac{1}{2}$	1	1	$1\frac{1}{2}$	2	2
3	1	1	$1\frac{1}{2}$	2	2
$3\frac{1}{2}$	1	1	$1\frac{1}{2}$	2	$2\frac{1}{2}$
4	1	1	$1\frac{1}{2}$	2	$2\frac{1}{2}$
$4\frac{1}{2}$	1	1	$1\frac{1}{2}$	2	$2\frac{1}{2}$
5	1	$1\frac{1}{2}$	$1\frac{1}{2}$	2	$2\frac{1}{2}$
6	1	$1\frac{1}{2}$	2	2	$2\frac{1}{2}$
7	$1\frac{1}{2}$	$1\frac{1}{2}$	2	2	$2\frac{1}{2}$
8	$1\frac{1}{2}$	$1\frac{1}{2}$	2	2	$2\frac{1}{2}$
9	$1\frac{1}{2}$	$1\frac{1}{2}$	2	$2\frac{1}{2}$	$2\frac{1}{2}$
10	$1\frac{1}{2}$	$1\frac{1}{2}$	2	$2\frac{1}{2}$	$2\frac{1}{2}$
12	$1\frac{1}{2}$	$1\frac{1}{2}$	2	$2\frac{1}{2}$	3
14 or more	$1\frac{1}{2}$	$1\frac{1}{2}$	2	$2\frac{1}{2}$	3

heat transmission, and the heat loss frequently becomes considerably greater than calculated. The reverse takes place with a pressure head on the cold side. Cold air then flows inwardly toward the increasingly hot brickwork and takes up heat that it transmits back to the interior. This action leads frequently to significant reduction of heat transfer, whereby the calculated heat losses by conduction exceed actual conditions. Thermal conductivity values that are too low can be explained by this effect.

The gas volume flowing through refractory walls depends on the porosity as well as the pressure head differential on both sides of the masonry. Careful investigation has shown that gas volume V passing through 1 sq ft of wall surface per hour is inversely proportional to wall thickness and increases with the differential pressure ΔP (in in. of water column) to a power between 1 and $\frac{1}{2}$. This reflects that part of the gas traverses the pores in laminar flow, when the power 1 is applicable, and another part flows turbulently through the larger cavities, when the radical should be introduced. Introducing as a first approximation the proportionality of gas flow rate to the differential pressure ΔP (in in. of water column), the following equation is developed:

$$V = \left(\frac{\varepsilon \, \Delta P}{L}\right)\left(\frac{\text{ft}^3}{\text{ft}^2\text{h}}\right) \tag{43}$$

The transmission coefficient ε varies from 2.75 to 55, depending on the state of the refractory. The value 2.75 applies to carefully built and slowly dried new walls. The value 55 is characteristic of walls with thin fissures, which appear after long operation at high temperatures, or when a wall has dried too quickly. The flow resistance increases proportionally to the mean absolute temperature of the refractory or gas. Therefore, ε decreases and becomes

$$\varepsilon = \varepsilon_o \frac{492}{T} \tag{43a}$$

Then, Equation 43 becomes

$$V_o = \left(\frac{\varepsilon_o \, \Delta P \, 492}{LT}\right) \frac{ft^3}{ft^2h} \tag{43b}$$

The rate of heat transmission through a wall is expressed by

$$Q = \frac{(t_1 - t_2)A}{L} (k \pm V \, Cp \, L) \text{ Btu/h} \tag{44}$$

In the preceding equations,

V	= gas volume, cu ft/h
ε	= transmission coefficient
ΔP	= pressure across wall, in. water column
T	= average gas temperature R°
L	= wall thickness, ft
h	= differential pressure, lb/sq ft
A	= surface area, sq ft
$(t_1 - t_2)$	= temperature gradient across wall, F°
k	= thermal conductivity, Btu/(ft)(h)(F°)
Cp	= specific heat, Btu/(cu ft)(F°)
Q	= heat transfer, Btu/h

The minus sign from Equation 44 applies when gas flows opposite to the direction of heat conduction. The plus sign is in the direction of heat conduction.

In the derivation of these equations, it is assumed that the temperature of the flowing gases is everywhere equal to the temperature of the wall because of the small size of the capillaries. This assumption is not always correct, however. If it were correct, the temperature of the outside surface of the boiler wall, with gases flowing past, producing a negative differential pressure, would always be equal to the temperature of ambient air. It would signify that the brickwork should not lose heat, which is not the case.

The temperature of air passing through the wall, in the first region, is below the temperature of the capillaries and fissures. During its passage, the air could attain the temperature of the wall at its inner side, and therefore would always be cooler than the heating gases flowing inside. The sucked air would then be mixed with the hotter gas masses, which would thus attain their final temperature. The heat required for this purpose is no longer a wall loss but a waste gas loss. The overall coefficient of heat transfer across a wall of refractory either of one material or several layers can be found by applying Equation 42.

A simplified method for determining required thickness of a refractory or insulating material is based on comparison with the equivalent thickness of firebrick, for example, the rough approximation that 1 in. of plastic refractory is equivalent to 2 in. of firebrick in insulating value. In this simplified method, each component or layer in a wall is converted to its equivalent thickness of firebrick. When the total equivalent thickness of the wall is known, the cold face temperature can be read directly from a curve; and with the cold face temperature, heat loss can be read from another curve.

The following brief outline presents the charts and curves used in this method.

Cold Face Temperature vs. Firebrick Wall Thickness

Figure 94 is used to find the cold face temperature for a given hot face temperature and thickness of firebrick. For example, with a $22\frac{1}{2}$-in. wall of firebrick and a 2200°F hot face, the cold face temperature is 316°F. This chart is based on 70°F air temperature, no wind, and vertical wall. These conditions are generally accepted as standard for heat transfer calculations.

Heat Loss from Vertical Walls

Figure 95 shows heat loss from vertical walls with various cold face temperatures and for differing conditions of wind velocity and surrounding air temperatures. For example, with no wind and surrounding air temperature of 70°F, a vertical wall with 200°F on the cold face will lose 290 Btu/(sq ft)(h). When Fig. 95 is used in conjunction with Fig. 94, the curve for 70°F air and no wind must be used.

Interface Temperature

Figure 96 is used to find interface temperatures—the temperature between two wall components. To use this chart, one must know heat loss through the wall (from Fig. 95), the hot face temperature, and the equivalent thickness of firebrick from the hot face to the interface. The following example is illustrated by Fig. 96. The hot face temperature is 2400°F, the equivalent

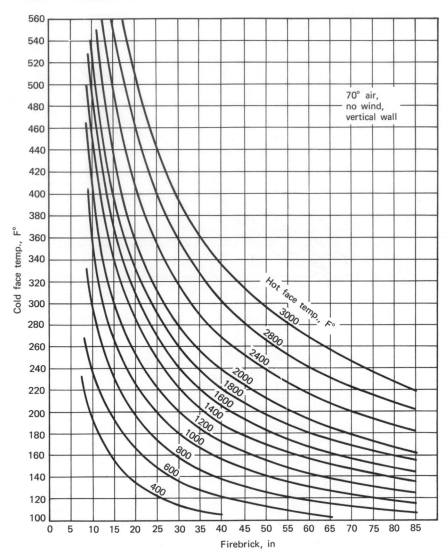

Fig. 94. Cold face temperature versus firebrick wall thickness.

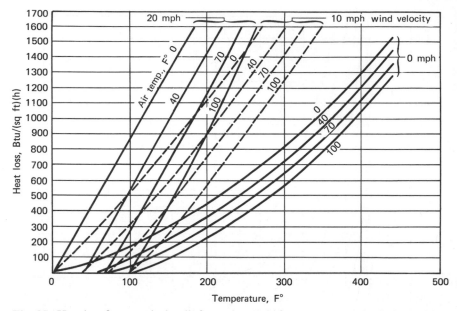

Fig. 95. Heat loss from vertical walls for various cold face temperatures, wind velocities, and air temperatures.

thickness of firebrick from the hot face to the interface is 19.80 in., and heat loss Q is 385 Btu/(h)(sq ft). Starting at the bottom-left edge, follow a line up from 19.80 in. until it intersects a line drawn horizontally from the left side at 385 Btu. From this point of intersection, follow a line at a 45° angle until it intersects the curve. (The 45° lines on the left half of this chart are for easy reference in determining this line.) From this point, follow horizontally to the right side of the chart until intersecting the line with the proper hot face temperature, 2400°F in this case. From this point, follow a vertical line down to the scale at the right side of the bottom edge. The interface temperature on this scale is 1650°F.

Equivalent Thickness of Firebrick for Various Temperatures and "K" Factors

Figure 97 illustrates equivalent in. of firebrick for various temperatures and "K" factors. By using this curve, the equivalent thickness for any material can be found if its "K" factor is known. While this method involves a trial-and-error approach, calculations are considerably reduced, with a good possibility of a correct first trial.

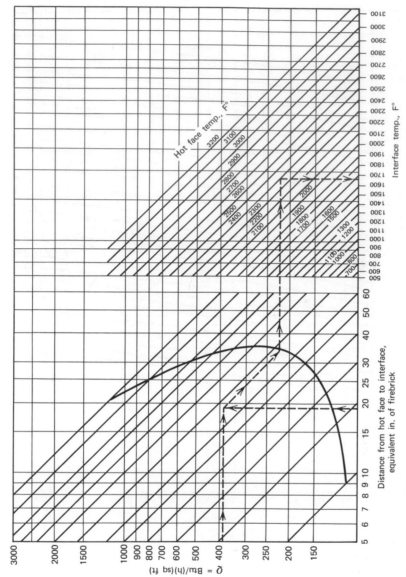

Fig. 96. Interface Temperature.

213

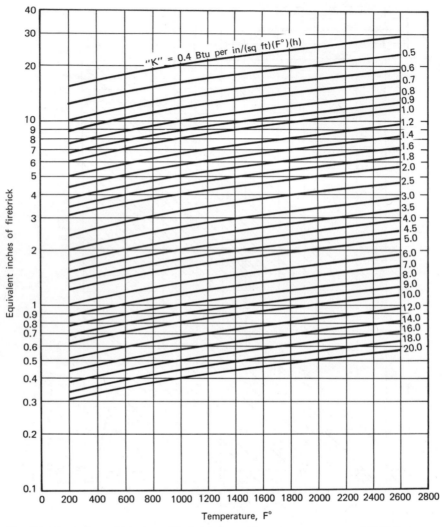

Fig. 97. Equivalent thickness of firebrick for various temperatures and "K" factors.

As stated, the foregoing is for vertical surfaces. Different curves apply for other surfaces. For surfaces facing up, shell temperatures are somewhat lower than for vertical surfaces. For surfaces facing down, shell temperatures are higher than for vertical surfaces.

The several types of available castable material have some common advantages: (1) rapid hardening, (2) virtually no drying shrinkage, (3) high resistance to spalling, and (4) negligible drying shrinkage in use at recommended temperatures. The adaptability of these materials makes possible the easy construction of monolithic furnaces, linings, and the like that would otherwise require the use of special shapes or involve costly and laborious cutting and fitting of brick. All grades of castable are as easy to handle as concrete. The choice of the castable depends on the particular service requirements involved. The highest service temperature for this material is 3000°F. It is high in alumina as a base material. Other materials are rated for 2600, 2400, and 2000°F and have an aluminum silicate base.

As an example of the method outlined for a composite wall, assume that from the hot side to the cold side the wall consists of

Six in. of castable, $k = 7.5$
Three in. of thermobestos blocks, $k = .60$
Four in. of 85% Magnesia, $k = .46$
One-quarter-in. of steel casing, $k = 300$

Hot side temperature—1400°F
Desired cold side temperature—130°F
U_1 hot side, from Figs. 22 and 23—60
U_2 cold side, from Figs. 22 and 23—050

From Equation 40, $U_o = \dfrac{1}{\dfrac{1}{.60} + \dfrac{6}{7.5} + \dfrac{3}{.60} + \dfrac{4}{.46} + \dfrac{.25}{300} + \dfrac{1}{.50}}$

$= 0.062$ Btu/(h)(sq ft)(F°)

From Equation 41, $q = U_o S \, \Delta t = (0.62)(1)(1400 - 130)$
$= 78.74$ Btu/(sq ft)(h)

From Equations 41a through 41e,

$t_0 - t_1 = 78.74/.60 \qquad = \quad 131°F \qquad$ (hot gas film)
$t_1 - t_2 = 78.74/(7.5/6) \quad = \quad 63°F \qquad$ (refractory)
$t_2 - t_3 = 78.74/(.60/3) \quad = \quad 393°F \qquad$ (Thermobestos)
$t_3 - t_4 = 78.74/(.46/4) \quad = \quad 685°F \qquad$ (85% Magnesia)
$t_4 - t_5 = 78.74/(300/.25) = \quad 1°F \qquad$ (steel casing)
$t_5 - t_6 = 78.74/.50 \qquad = \quad 157°F \qquad$ (cold air film)

Total $\Delta T = 1430°F$

Cold face temperature $= 1400 - 1273 = 127°F$

The target cold side surface temperature was assumed to be 130°F. Since the results of the first trial bear this out, an additional trial is not required. The graphical method described is obvious in its use and accordingly will not be further elaborated. It is recommended that suppliers of insulating and refractory materials be consulted regarding physical characteristics of the material.

APPENDIX A

THERMAL CONDUCTIVITY TABLES

TABLE A-1 Thermal Conductivity of Commonly Used Materials

$$[k = \text{Btu}/(\text{h})(\text{sq ft})(\text{F}°)/(\text{ft})]$$

Material	Temperature (F°)	k	Material	Temperature (F°)	k
Metals			*Metals*		
Aluminum	212	119	Hasteloy X	1100	12.0
Aluminum	392	124	Hasteloy X	1500	14.5
Aluminum	752	144	Hasteloy X	1700	15.7
Aluminum brass	68	58	Inconel 625	1000	10.1
Aluminum			Inconel 625	1600	13.2
bronze 5%	68	46	Inconel 625	1800	14.6
Admiralty metal	86	65	Inconel 702	400	10.4
Brass (70 Cu-30			Inconel 702	800	15.0
Zn)	212	60	Inconel 702	1200	19.6
Brass (70 Cu-30			Inconel X750	400	8.2
Zn)	392	63	Inconel X750	800	10.0
Brass (70 Cu-30			Inconel X750	1600	13.7
Zn)	752	67	Incoloy 800	1000	11.6
Cupro-nickel 10%	68	26	Incoloy 800	1400	13.8
Cupro-nickel 20%	68	21	Incoloy 800	1800	17.8
Cupro-nickel 30%	68	17	Incoloy 825	800	9.9
Cadmium	64	53.7	Incoloy 825	1000	10.9
Cadmium	212	52.2	Incoloy 825	1400	12.9
Cast iron	212	30.0	Incoloy 825	1800	16.0
Cast iron	392	28	Lead	212	19.0
Cast iron	752	25	Lead	392	18.0
Copper (pure)	212	218	Monel 400	200	13.9
Copper (pure)	392	215	Monel 400	600	17.9
Copper (pure)	752	210	Monel 400	1400	25.9
Graphite	212	87	Monel K-500	400	13.0
Graphite	752	58	Monel K-500	800	15.7
Hasteloy 25	900	10.6	Monel K-500	1600	23.5
Hasteloy 25	1300	11.9	Nickel	212	34.0
Hasteloy 25	1700	15.9	Nickel	392	33.0

TABLE A-1 (*continued*)

Material	Temperature (F°)	k	Material	Temperature (F°)	k
Metals			*Insulation and*		
Nickel	572	32.0	*Refractory*		
Silver	212	238.0	Diatom. earth-		
Steel (mild)	212	26.0	brick (1600)	600	0.055
Steel (mild)	392	26.0	Diatom. earth-		
Steel (mild)	572	25.0	brick (1600)	1000	0.065
Steel (stainless)			Diatom. earth-		
301, 302, 303,			brick (2000)	400	0.137
304, 316	212	9.4	Diatom. earth-		
Steel (stainless)			brick (2000)	600	0.140
301, 302, 303,			Diatom. earth-		
304, 316	932	12.4	brick (2000)	1000	0.158
Steel (stainless)			Diatom. earth-		
308	212	8.8	brick (2500)	600	0.148
Steel (stainless)			Diatom. earth-		
308	932	12.5	brick (2500)	1000	0.163
Steel (stainless)			Diatom. earth-		
309, 310	212	8.0	brick (2500)	2000	0.203
Steel (stainless)			Fireclay	392	0.580
309, 310	932	10.8	Kaolin ins. brick	932	0.150
Steel (stainless)			Kaolin ins. brick	2100	0.260
321, 347	212	9.3	Magnesite	399	2.20
Steel (stainless)			Magnesite	1200	1.60
321, 347	932	12.8	Magnesite	2192	1.10
			Magnesia 85%	200	0.036
			Magnesia 85%	300	0.038
			Magnesia 85%	400	0.040
Insulation and			Molded pipe		
Refractory			covering	400	0.051
Asbestos sheets	124	0.096	Molded pipe		
Brick (Alumina			covering	1600	0.088
92–99)	800	1.80	Rock wool	200	0.034
Brick (Alumina			Rock wool	400	0.044
64–65)	2400	2.70	Rock wool	600	0.057
Brick (chrome)	392	0.67	Silicon carbide		
Brick (chrome)	1200	0.85	brick	1112	10.70
Brick (chrome)	2400	1.00	Silicon carbide		
Corrugated			brick	1475	9.20
asbestos	200	0.058	Silicon carbide		
Corrugated			brick	1832	8.00
asbestos	300	0.069	Silicon carbide		
Diatom. earth-			brick	2552	6.30
brick (1600)	400	0.050			

TABLE A-1   (*continued*)

Material	Temperature (F°)	k	Material	Temperature (F°)	k
Liquids			*Liquids*		
Ammonia	5–86	0.29	Water	300	0.395
Alcohol	86	0.097	Water	420	0.376
Alcohol	167	0.095	Water	620	0.275
Calcium chloride					
brine	86	0.32	*Gases and Vapors*		
Dichlorodifluoro-			Acetylene	−103	0.0068
methane	20	0.057	Acetylene	32	0.0108
Dichlorodifluoro-			Acetylene	122	0.0140
methane	60	0.053	Acetylene	212	0.0172
Dichlorodifluoro-			Air	−328	0.0040
methane	100	0.048	Air	−148	0.0091
Dichlorodifluoro-			Air	32	0.0140
methane	180	0.038	Air	212	0.0184
Ethylene glycol	32	0.153	Air	392	0.0224
Gasoline	86	0.078	Air	572	0.0260
Glycerol 100%	68	0.164	Ammonia	−58	0.0097
Glycerol 80%	68	0.189	Ammonia	32	0.0126
Glycerol 60%	68	0.220	Ammonia	212	0.0192
Glycerol 40%	68	0.259	Ammonia	392	0.0280
Glycerol 20%	68	0.278	Ammonia	572	0.0385
Glycerol 100%	212	0.164	Ammonia	752	0.0509
Kerosene	68	0.086	Butane	32	0.0078
Kerosene	167	0.081	Butane	212	0.0135
Mercury	82	4.83	Carbon dioxide	−58	0.0064
Methyl alcohol			Carbon dioxide	32	0.0084
100%	68	0.124	Carbon dioxide	212	0.0128
Methyl alcohol			Carbon dioxide	392	0.0177
80%	68	0.154	Carbon dioxide	572	0.0229
Methyl alcohol			Carbon monoxide	−328	0.0037
60%	68	0.190	Carbon monoxide	−148	0.0088
Methyl alcohol			Carbon monoxide	32	0.0134
40%	68	0.234	Carbon monoxide	212	0.0176
Methyl alcohol			Chlorine	32	0.0043
20%	68	0.284	Dichlorodifluoro-		
Methyl alcohol			methane	32	0.0048
100%	122	0.114	Dichlorodifluoro-		
Oils	86	0.079	methane	122	0.0064
Trichloroethylene	122	0.080	Dichlorodifluoro-		
Water	32	0.343	methane	212	0.0080
Water	100	0.363	Dichlorodifluoro-		
Water	200	0.393	methane	302	0.0097

TABLE A-1   *(continued)*

Material	Temperature (F°)	k	Material	Temperature (F°)	k
Gases and Vapors			*Gases and Vapors*		
Ethyl alcohol	68	0.0089	Nitrogen	212	0.0181
Ethyl alcohol	212	0.0124	Nitrogen	392	0.0220
Ethylene	−96	0.0064	Nitrogen	572	0.0255
Ethylene	32	0.0101	Nitrogen	752	0.0287
Ethylene	122	0.0131	Nitrous oxide	−148	0.0047
Ethylene	212	0.0161	Nitrous oxide	32	0.0088
Helium	−328	0.0338	Nitrous oxide	212	0.0138
Helium	−148	0.0612	Oxygen	−328	0.0038
Helium	32	0.0818	Oxygen	−148	0.0091
Helium	212	0.0988	Oxygen	32	0.0142
Hydrogen	−328	0.0293	Oxygen	122	0.0166
Hydrogen	−148	0.0652	Oxygen	212	0.0188
Hydrogen	32	0.0966	Propane	32	0.0074
Hydrogen	212	0.1240	Propane	212	0.0151
Hydrogen	392	0.1484	Sulfur dioxide	32	0.0050
Hydrogen	572	0.1705	Sulfur dioxide	212	0.0069
Mercury	392	0.0197	Water vapor		
Methane	−328	0.0045	(zero press.)	212	0.0136
Methane	−148	0.0109	Water vapor		
Methane	32	0.0176	(zero press.)	392	0.0182
Methane	212	0.0255	Water vapor		
Methane	392	0.0358	(zero press.)	572	0.0230
Methane	572	0.0490	Water vapor		
Nitrogen	−328	0.0040	(zero press.)	752	0.0279
Nitrogen	−148	0.0091	Water vapor		
Nitrogen	32	0.0139	(zero press.)	932	0.0328

VISCOSITY TABLES AND CURVES

TABLE B-1 Viscosities of Commonly Used Materials

(centipoises $[\mu]$)

Material	Temperature (F°)	μ	Material	Temperature (F°)	μ
Liquids			*Liquids*		
Ammonia 100%	−20	0.230	Freon-11	40	0.55
Ammonia 100%	−10	0.215	Freon-12	−20	0.38
Ammonia 100%	0	0.200	Freon-12	−10	0.37
Ammonia 100%	10	0.180	Freon-12	0	0.35
Ammonia 100%	20	0.160	Freon-12	10	0.34
Ammonia 100%	30	0.150	Freon-12	40	0.29
Ammonia 100%	40	0.140	Freon-21	−20	0.55
Ammonia 100%	80	0.100	Freon-21	−10	0.50
Brine, CaCl$_2$ 25%	0	9.50	Freon-21	0	0.49
Brine, CaCl$_2$ 25%	30	5.00	Freon-21	10	0.47
Brine, CaCl$_2$ 25%	60	2.90	Freon-21	20	0.45
Brine, CaCl$_2$ 25%	100	1.40	Freon-21	40	0.41
Brine, CaCl$_2$ 25%	150	0.63	Freon-22	−20	0.31
Brine, CaCl$_2$ 25%	200	0.29	Freon-22	−10	0.30
Brine, NaCl 25%	0	5.00	Freon-22	0	0.29
Brine, NaCl 25%	30	3.40	Freon-22	10	0.28
Brine, NaCl 25%	60	2.50	Freon-22	40	0.25
Brine, NaCl 25%	100	1.60	Freon-113	−20	1.50
Brine, NaCl 25%	150	0.95	Freon-113	−10	1.35
Brine, NaCl 25%	200	0.62	Freon-113	0	1.22
Ethylene glycol	0	120	Freon-113	10	1.15
Ethylene glycol	50	35	Freon-113	40	0.90
Ethylene glycol	100	13	Glycerol 100%	150	75.00
Ethylene glycol	150	5.3	Glycerol 100%	200	18.00
Ethylene glycol	200	2.4	Glycerol 100%	300	1.50
Freon-11	−20	0.8	Glycerol 50%	0	22.00
Freon-11	−10	0.75	Glycerol 50%	100	3.60
Freon-11	0	0.70	Glycerol 50%	200	0.80
Freon-11	10	0.65	Glycerol 50%	300	0.22

TABLE B-1 (continued)

Material	Temperature (F°)	μ	Material	Temperature (F°)	μ
Liquids			*Gases*		
Kerosene	0	5.00	Ammonia	200	0.0125
Kerosene	50	2.90	Ammonia	300	0.0144
Kerosene	100	1.65	Butane	40	0.0078
Kerosene	150	1.00	Butane	212	0.0100
Kerosene	200	0.65	Carbon dioxide	−100	0.0100
Mercury	0	1.85	Carbon dioxide	0	0.0125
Mercury	50	1.65	Carbon dioxide	100	0.0148
Mercury	200	1.25	Carbon dioxide	200	0.0170
Mercury	300	1.10	Carbon dioxide	400	0.0215
Mercury	350	1.00	Carbon dioxide	1000	0.0345
Sodium	0	1.40	Carbon		
Sodium	50	1.15	monoxide	−100	0.0133
Sodium	100	0.96	Carbon		
Sodium	150	0.82	monoxide	0	0.0160
Sodium	200	0.72	Carbon		
Sodium	300	0.55	monoxide	100	0.0182
Sulfuric acid 60%	0	15.00	Carbon		
Sulfuric acid 60%	100	4.60	monoxide	200	0.0205
Sulfuric acid 60%	200	1.80	Carbon		
Sulfuric acid 60%	300	0.82	monoxide	400	0.0250
Trichloroethylene	0	0.88	Carbon		
Trichloroethylene	100	0.52	monoxide	1000	0.0360
Trichloroethylene	200	0.33	Freon-11	−100	0.0080
Water	50	1.20	Freon-11	0	0.0096
Water	100	0.72	Freon-11	40	0.0103
Water	200	0.28	Freon-11	100	0.0111
			Freon-11	200	0.0128
Gases			Freon-12	−100	0.0093
Acelylene	0	0.0065	Freon-12	0	0.0110
Acelylene	100	0.0080	Freon-12	40	0.0118
Acelylene	200	0.0096	Freon-12	100	0.0125
Acelylene	300	0.0114	Freon-12	200	0.0142
Air	−100	0.0132	Freon-21	−100	0.0085
Air	0	0.0160	Freon-21	0	0.0102
Air	100	0.0180	Freon-21	40	0.0108
Air	200	0.0205	Freon-21	100	0.0118
Air	300	0.0230	Freon-21	200	0.0132
Ammonia	−50	0.0078	Freon-22	−100	0.0094
Ammonia	0	0.0086	Freon-22	0	0.0112
Ammonia	100	0.0105	Freon-22	40	0.0120

TABLE B-1   (*continued*)

Material	Temperature (F°)	μ	Material	Temperature (F°)	μ
Gases			*Gases*		
Freon-22	100	0.0132	Methane	200	0.0140
Freon-22	200	0.0150	Methane	300	0.0158
Freon-113	−100	0.0080	Methane	400	0.0172
Freon-113	0	0.0093	Nitrogen	−100	0.0130
Freon-113	40	0.0100	Nitrogen	0	0.0156
Freon-113	100	0.0107	Nitrogen	100	0.0180
Freon-113	200	0.0118	Nitrogen	200	0.0202
Chlorine	32	0.0122	Nitrogen	500	0.0275
Chlorine	100	0.0140	Nitrogen	1000	0.0370
Ethylene	0	0.0087	Oxygen	−100	0.0150
Ethylene	100	0.0102	Oxygen	0	0.0180
Ethylene	200	0.0120	Oxygen	100	0.0210
Helium	0	0.0169	Oxygen	200	0.0230
Helium	100	0.0192	Oxygen	500	0.0305
Helium	200	0.0220	Oxygen	1000	0.0410
Helium	300	0.0240	Propane	−100	0.0056
Hydrogen	0	0.0080	Propane	0	0.0070
Hydrogen	100	0.0090	Propane	100	0.0082
Hydrogen	200	0.0100	Sulfur dioxide	0	0.0108
Hydrogen	300	0.0110	Sulfur dioxide	100	0.0126
Mercury	0	0.0140	Sulfur dioxide	200	0.0145
Mercury	100	0.0190	Water	100	0.0100
Mercury	200	0.0250	Water	200	0.0120
Mercury	400	0.0380	Water	400	0.0162
Methane	0	0.0103	Water	600	0.0205
Methane	100	0.0120	Water	700	0.0230

APPENDIX C

VISCOSITY CONVERSION FORMULAE

TABLE C-1 Viscosity Conversion Formulae

Viscosity in centipoises $\div$ 100 = poises = gm/(sec)(cm)
Viscosity in centipoises $\times$ 0.000672 = lb/(sec)(ft)
Viscosity in centipoises $\times$ 0.0000209 = lb force/(sec)/(sq ft)
Viscosity in centipoises $\times$ 2.42 = lb/(h)(ft) = absolute viscosity
Viscosity in centipoises $\times$ 3.60 = kg/(h)(m)

APPENDIX D

DENSITY TABLES

TABLE D-1 Density

$$(\text{lb/cu ft} = \rho)$$

Material	ρ	Material	ρ
Element		*Element*	
Aluminum	168.6	Tungsten	1206
Antimony	413.0	Vanadium	350
Barium	225.0	Zinc	446
Bismuth	612.0		
Cadmium	540.0	*Alloys and Other Materials*	
Calcium	97.0	Aluminum bronze	481
Carbon	141.0	Brass	534
Chromium	446.0	Bronze	509
Cobalt	550.0	Bronze (phosphor)	554
Copper	558.0	Cast iron	442
Gold	1206.0	Iron, wrought	485
Hydrogen	53×10^{-4}	Monel metal	555
Iron	492	Steel (cold drawn)	489
Lead	708		
Magnesium	108		
Manganese	462	*Liquids*	
Mercury	846	Alcohol, ethyl	49
Molybdenum	637	Alcohol, methyl	50
Nickel	556	Acid, muriatic (40%)	75
Nitrogen	73×10^{-3}	Acid, nitric (91%)	94
Oxygen	83×10^{-3}	Acid, sulfuric (87%)	112
Phosphorus	114	Ether	46
Platinum	1340	Oils, vegetable	58
Potassium	54.7	Turpentine	54
Silicon	145	Water, 100°C	62.428
Silver	656		
Sodium	61		
Sulphur	129	*Minerals*	
Tin	450	Asbestos	153
Titanium	281	Basalt	184

TABLE D-1 (*continued*)

Material	ρ		
Minerals		*Bituminous Substances*	
Bauxite	159	Asphaltum	81
Borax	109	Coal, anthracite	97
Clay	137	Coal, bituminous	84
Dolomite	181	Coal, lignite	78
Granite	165	Coal, peat, turf, dry	47
Gypsum	159	Coal, charcoal, pine	23
Limestone	155	Coal, charcoal, oak	33
Marble	170	Coal, coke	75
Magnesite	187	Graphite	135
Pumice	40	Paraffin	56
Quartz	165	Petroleum	54
Sandstone	143	Petroleum (kerosene)	50
Shale	172	Petroleum (gasoline)	45
Soapstone	169	Pitch	69

SPECIFIC HEAT TABLES

TABLE E-1   Specific Heat of Commonly Used Materials

$$[Cp = Btu/(lb)(F°) = gm\text{-}cal/(gm)(C°)]$$

Material	Temperature (F°)	Cp	Material	Temperature (F°)	Cp
Metals			*Metals*		
Lead	32	0.0306	Quartz	212	0.2061
Lead	212	0.0315	Quartz	392	0.2315
Lead	392	0.0325			
Zinc	32	0.0917			
Zinc	212	0.0958	*Liquids*		
Zinc	392	0.1000	Ammonia	−200	1.100
Aluminum	32	0.2106	Ammonia	−100	1.400
Aluminum	212	0.2225	Brine, 25% $CaCl_2$	−50	0.640
Aluminum	392	0.2344	Brine, 25% $CaCl_2$	0	0.660
Silver	32	0.0557	Brine, 25% $CaCl_2$	50	0.680
Silver	212	0.0571	Brine, 25% $CaCl_2$	100	0.700
Silver	392	0.0585	Brine, 25% NaCl	0	0.800
Gold	32	0.0305	Brine, 25% NaCl	50	0.810
Gold	212	0.0312	Brine, 25% NaCl	100	0.820
Gold	392	0.0320	Carbon tetra-		
Coper	32	0.0919	chloride	0	0.190
Copper	212	0.0942	Carbon tetra-		
Copper	392	0.0965	chloride	100	0.210
Nickel	32	0.1025	Carbon tetra-		
Nickel	212	0.1132	chloride	200	0.230
Nickel	392	0.1241	Dowtherm A	100	0.400
Iron	32	0.1051	Dowtherm A	200	0.450
Iron	212	0.1166	Dowtherm A	300	0.510
Iron	392	0.1280	Dowtherm A	400	0.560
Cobalt	32	0.1023	Ethyl alcohol		
Cobalt	212	0.1079	100%	0	0.4350
Cobalt	392	0.1138	Ethyl alcohol		
Quartz	32	0.1667	100%	100	0.6350

TABLE E-1 *(continued)*

Material	Temperature (F°)	C_p	Material	Temperature (F°)	C_p
Liquids			*Liquids*		
Ethyl alcohol					
50%	0	0.830	Sulfur dioxide	−20	0.300
Ethyl alcohol			Sulfur dioxide	0	0.330
50%	100	0.940	Sulfur dioxide	100	0.350
Ethylene glycol	−50	0.500	Water	32	1.000
Ethylene glycol	0	0.540	Water	100	1.050
Ethylene glycol	50	0.565	Water	212	1.080
Ethylene glycol	200	0.665			
Freon-11	−20	0.200	*Gases at 1 atm*		
Freon-11	0	0.210	Air	32	0.245
Freon-11	100	0.220	Air	100	0.250
Freon-12	−40	0.220	Air	200	0.255
Freon-12	0	0.230	Air	500	0.260
Freon-12	100	0.270	Air	1000	0.265
Freon-21	−20	0.245	Ammonia	0	0.500
Freon-21	0	0.250	Ammonia	50	0.520
Freon-21	100	0.255	Ammonia	100	0.525
Freon-22	−20	0.260	Ammonia	500	0.600
Freon-22	0	0.270	Carbon dioxide	0	0.205
Freon-22	100	0.310	Carbon dioxide	100	0.215
Freon-113	−20	0.205	Carbon dioxide	500	0.250
Freon-113	0	0.210	Carbon dioxide	1000	0.280
Freon-113	100	0.220	Carbon dioxide	1500	0.300
Gasoline	32–212	0.50	Carbon monoxide	0	0.250
Glycerol	−40	0.5000	Carbon monoxide	100	0.255
Glycerol	0	0.5300	Carbon monoxide	500	0.260
Glycerol	50	0.5600	Carbon monoxide	1000	0.270
Methyl alcohol	−40	0.540	Chlorine	100	0.115
Methyl alcohol	0	0.565	Chlorine	500	0.127
Methyl alcohol	50	0.590	Chlorine	1000	0.140
Methyl chloride	−80	0.350	Ethylene	100	0.400
Methyl chloride	0	0.370	Ethylene	200	0.440
Methyl chloride	50	0.380	Ethylene	300	0.500
Pyridine	−50	0.370	Ethylene	400	0.550
Pyridine	0	0.385	Freon-11	0	0.128
Pyridine	100	0.420	Freon-11	100	0.140
Pyridine	200	0.450	Freon-11	200	0.150
Sulfuric acid 98%	0	0.330	Freon-21	0	0.140
Sulfuric acid 98%	50	0.340	Freon-21	100	0.150
Sulfuric acid 98%	100	0.355	Freon-21	200	0.162

TABLE E-1 (*continued*)

Material	Temperature (F°)	Cp	Material	Temperature (F°)	Cp
Gases at 1 atm			*Gases at 1 atm*		
Freon-22	0	0.150	Nitrogen	500	0.260
Freon-22	100	0.161	Nitrogen	1000	0.270
Freon-22	200	0.173	Oxygen	0	0.215
Freon-113	0	0.151	Oxygen	100	0.220
Freon-113	100	0.161	Oxygen	200	0.225
Freon-113	200	0.170	Oxygen	500	0.235
Hydrogen	0	3.450	Oxygen	1000	0.255
Hydrogen	100	3.500	Sulfur dioxide	0	0.150
Hydrogen	500	3.510	Sulfur dioxide	200	0.162
Methane	0	0.510	Sulfur dioxide	400	0.185
Methane	100	0.550	Sulfur dioxide	600	0.188
Methane	300	0.630	Sulfur dioxide	1000	0.212
Methane	500	0.725	Water	0	0.435
Methane	1000	0.940	Water	200	0.450
Nitrogen	0	0.250	Water	500	0.465
Nitrogen	100	0.255	Water	1000	0.500
Nitrogen	200	0.257	Water	1500	0.550

EXHAUST CHARACTERISTICS OF SOME GAS TURBINES

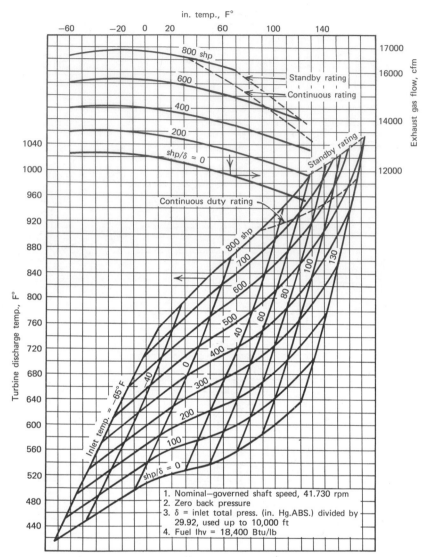

Fig. F-1. Gas turbine performance curve for Airesearch model IE831–800.

in. temp., F°

Exhaust gas flow, cfm

Turbine discharge temp., F°

800 shp

600

400

200

shp/δ = 0

Standby rating

Continuous rating

Standby rating

Continuous duty rating

800 shp

700

600

500

400

300

200

100

shp/δ = 0

130

100

80

60

40

0

40

Inlet temp. = −65° F

1. Nominal—governed shaft speed, 41.730 rpm
2. Zero back pressure
3. δ = inlet total press. (in. Hg.ABS.) divided by 29.92, used up to 10,000 ft
4. Fuel lhv = 18,400 Btu/lb

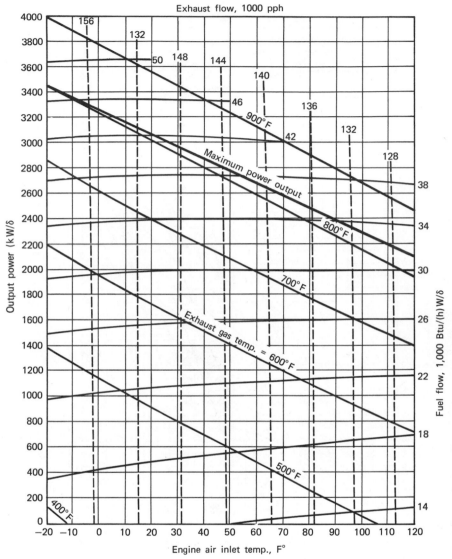

Fig. F-2. Solar Centaur continuous duty gas turbine generator set. Zero inlet and exhaust pressure losses.

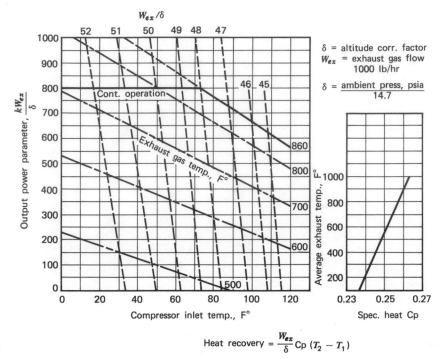

Fig. F-3. Available exhaust heat recovery from Solar Saturn continuous duty gas turbine generator set.

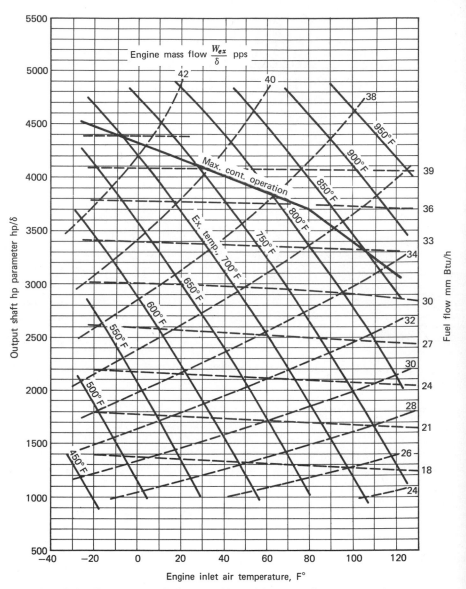

Fig. F-4. Solar Centaur two-shaft gas turbine minimum performance, optimum power. Zero gearbox and external duct losses.

EXHAUST CHARACTERISTICS OF SOME RECIPROCATING GAS AND DIESEL ENGINES

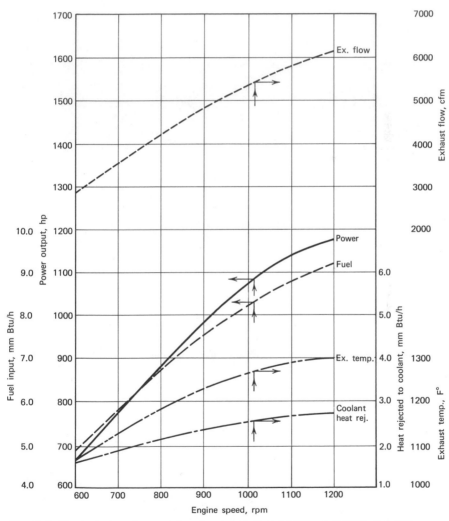

Fig. G-1. Typical natural gas engine characteristics at full load. 8 CYL. 8:1 CR.

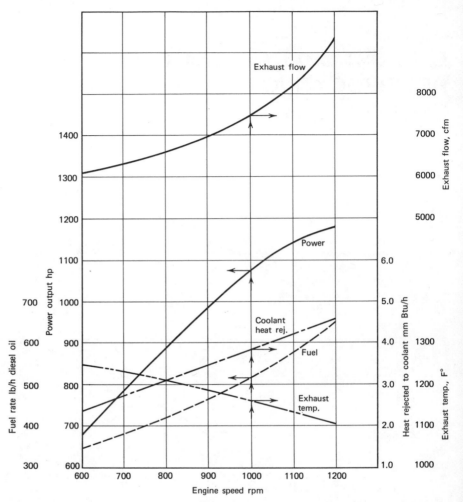

Fig. G-2. Typical turbo-charged intercooled diesel engine, characteristics at full load.

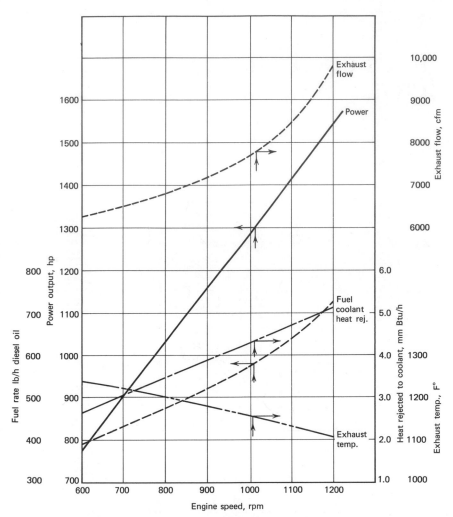

Fig. G-3. Typical turbo-charged diesel engine, characteristics at full load.

CROSS-COUNTERFLOW CORRECTION CURVES

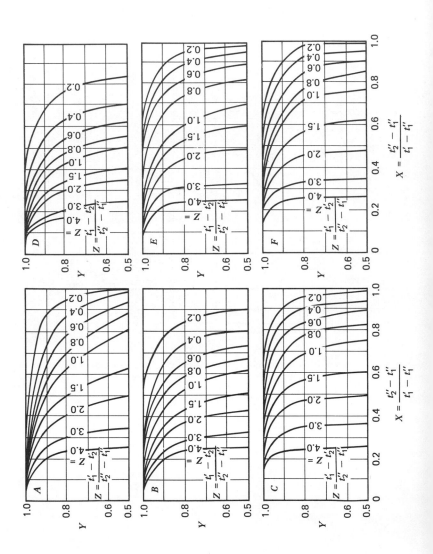

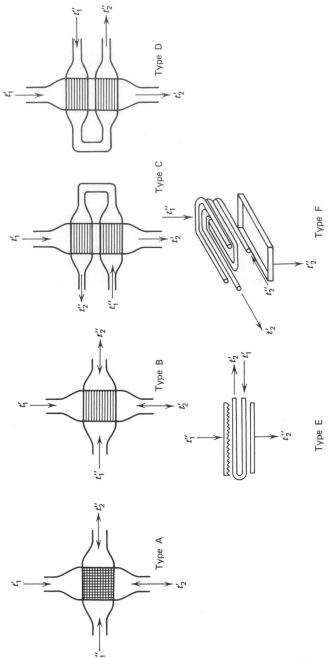

Fig. H. Correction factors for logarithmic mean temperature difference, cross flow heat exchangers.

APPENDIX I

CHART—RADIATION FROM FLAME TO COMBUSTION CHAMBER WALLS AND ABSORPTION SURFACES

Example: assume that it is required to find the heat transfer rate by radiation from a combustion flame within a combustion chamber having a partial water wall. The combustion chamber is 4 ft in diameter and 10 ft long. The fuel oil firing rate is 2137 lb/h. The effective surface subject to radiation heat transfer is 60 sq ft. Estimated combustion temperature is 3000°F (from excess air and fuel data). The surface temperature of the radiant heat receiving surface is 400°F.

1. High heating value of fuel—19,400 Btu/lb
2. Heat of combustion (19,400)(2137) = 41,457,800 Btu/h
3. Total surface (including one end) of combustion chamber = 138.23 sq ft
4. Heat release rate = 41,457,800/138.23 = 300,000 Btu/(sq ft)(h)
5. On Fig. A-6, draw line *A-B* from temperature 3000°F to heat release rate 300,000 Btu/(sq ft)(h)
6. Where line intersects the 400°F curve at *C*, draw a horizontal line to *D*. Read 76,000 Btu per sq ft
7. Multiply 76,000 by the effective surface subject to radiation (60 sq ft). The answer is 4,560,000 Btu/h, which is the rate of radiant heat transfer under the conditions described above.

The preceding method may be used to estimate direct radiation heat transfer.

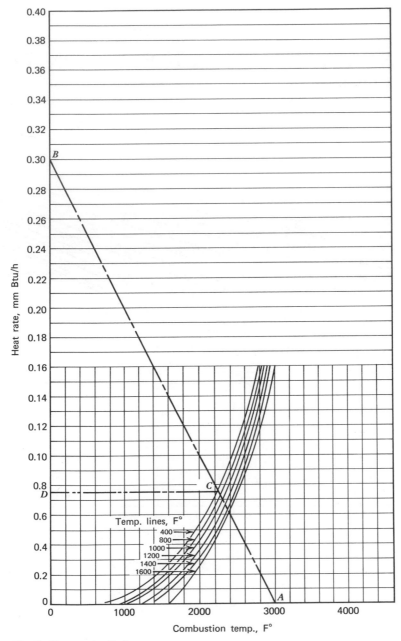

Fig. I. Furnace radiation.

CURVE OF RADIATION BETWEEN TWO BODIES

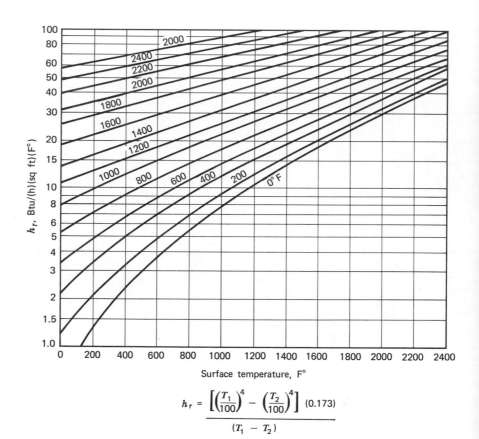

Surface temperature, F°

$$h_r = \frac{\left[\left(\frac{T_1}{100}\right)^4 - \left(\frac{T_2}{100}\right)^4\right](0.173)}{(T_1 - T_2)}$$

Emissivity $E = 1.0$; T = deg. abs.
Numbers on curve refer to temperature
of other surface

Fig. J. Coefficient of heat transfer by radiation for emissivity $E = 1.0$.

CHART OF DUCT RESISTANCE

Figure K is used to determine the resistance of air flowing in ducts. If air flow and duct diameter are known, proceed from the left to right to proper duct size line. Then proceed down and read duct resistance or pressure drop. The diagonal lines read air velocity.

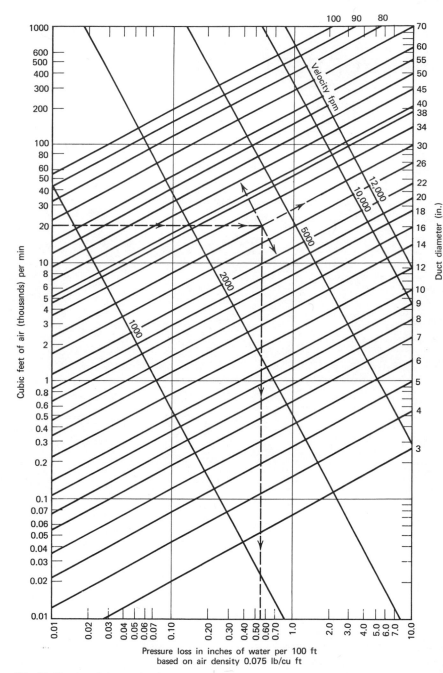

Fig. K Duct resistance chart.

APPENDIX L

NOMENCLATURE

A = Total effective heat transfer surface, sq ft
Cp = Specific heat at constant pressure, Btu/(lb)(F°)
D = Inside or outside tube diameter, ft
d = Inside or outside tube diameter, in.
d_c = Diameter of coil, in.
G = Mass velocity, lb/(h)(sq ft) of cross section
G_1 = Mass velocity, lb/(sec)(sq ft) of cross section
g = Acceleration of gravity, 4.18×10^8 ft/h²
h = Film coefficient of convection, Btu/(h)(sq ft)(F°)
h_s = Scale coefficient, Btu/(h)(sq ft)(F°)
k = Thermal conductivity, Btu/(h)(ft)(F°)
N = Length of tube, or height of surface, ft
n = Number of tubes directly over each other
P = Pressure, lb/sq ft absolute
p = Pressure, atm. absolute
R = Gas constant, ft per F°
T = Temperature, F° absolute
t = Temperature, F°
Δt = Temperature difference, F°
Lm ΔT = Log-mean temperature difference, F°
V = Linear velocity, ft per h
V_1 = Linear velocity, ft per sec
v = Specific volume, cu ft per lb
W = Rate of condensation or evaporation, lb/(sq ft)(h)
W^1 = Rate of flow, condensed or evaporated, lb/h per tube
      Thermal coefficient of expansion, in. per in./F°
      Absolute Viscosity, lb/(ft)(h)
      Density, lb/(cu ft)(h)
q = Rate of heat transfer, Btu/h
L = Thickness of body, ft
U = Overall coefficient of heat transfer, Btu/(h)(sq ft)(F°)
η = Efficiency
ρ = Density, lb/cu ft

BIBLIOGRAPHY ————————————————

1. *Applied Heat Transmission*, Herman J. Stoever, New York McGraw-Hill Book Co., 1941.

2. *Heat Transmission*, 3rd Edition, William H. McAdams, New York, McGraw-Hill Book Co., 1954.

3. *Industrial Heat Transfer*, 6th Edition, Alfred Schack, New York, John Wiley & Sons, 1965.

4. *Steam, Its Generation and Use*, 38th Edition, Babcock & Wilcox Co., New York, 1972.

5. *Thermodynamic Properties of Steam*, 20th Printing, Joseph H. Keenen and Frederick G. Keyes, New York, John Wiley & Sons, 1949.

6. *Thermodynamic Properties of Air*, Joseph H. Keenan and Joseph Kay, New York, John Wiley & Sons, 1945.

7. *Comparing Combined Cycle Plants*, Leroy O. Tomlinson, *Gas Turbine International*, p. 20–27, November–December, 1972.

8. *Gas Turbine Heat Recovery at California State Mall*, John L. Boyen, *Diesel and Gas Turbine Progress*, p. 54–55, February, 1970.

9. *Shell Platform Utilizes Unique Heat Recovery System*, John L. Boyen, *Petroleum Engineer*, p. 60–62, 64, April, 1970.

10. *Steam Injection, A Source of Incremental Power*, Charles Bultzo, American Society of Mechanical Engineers Publication 69-GT-68, 1968.

CONTRIBUTORS ──────────────

1. Airesearch Mfg. Corporation
2. American Schack Company, Inc.
3. Babcock & Wilcox Company
4. Conseco, Inc.
5. Dow Chemical Company
6. Exxon Corporation
7. Escoa Fin-Tube Corporation
8. Henry Vogt Machine Comapny
9. Hutchison-Hayes International
10. Monsanto Industrial Chemicals Company
11. Struthers Wells Corporation
12. Solar, Division of International Harvester Co.
13. Union Carbide Corporation

INDEX